Note Taking Guide

Precalculus with Limits
A Graphing Approach

TEXAS EDITION

Ron Larson
The Pennsylvania State University,
The Behrend College

Prepared by

Emily Keaton

Australia • Brazil • Mexico • Singapore • United Kingdom • United States

For product information and technology assistance, contact us at **Cengage Learning Customer & Sales Support, 1-800-354-9706**.

For permission to use material from this text or product, submit all requests online at **www.cengage.com/permissions**
Further permissions questions can be emailed to **permissionrequest@cengage.com**.

ISBN-13: 978-1-285-86779-3
ISBN-10: 1-285-86779-3

Cengage Learning
200 First Stamford Place, 4th Floor
Stamford, CT 06902
USA

Cengage Learning is a leading provider of customized learning solutions with office locations around the globe, including Singapore, the United Kingdom, Australia, Mexico, Brazil, and Japan. Locate your local office at: **www.cengage.com/global**.

Cengage Learning products are represented in Canada by Nelson Education, Ltd.

To learn more about Cengage Learning Solutions, visit **www.cengage.com**.

Purchase any of our products at your local college store or at our preferred online store **www.cengagebrain.com**.

Printed in the United States of America
2 3 4 5 6 7 18 17 16 15

Table of Contents

Table of Contents

Name ______________________________ Date __________

Chapter 1 Functions and Their Graphs

Section 1.1 Lines in the Plane

Objective: In this lesson you learned how to find and use the slope of a line to write and graph linear equations.

Important Vocabulary	Define each term or concept.
Slope	
Parallel	
Perpendicular	

I. The Slope of a Line (Pages 3–4)

> ***What you should learn***
> How to find the slopes of lines

The formula for the **slope** of a line passing through the points (x_1, y_1) and (x_2, y_2) is $m =$ ______________________ .

To find the slope of the line through the points $(-2, 5)$ and $(4, -3)$, ____

____.

A line whose slope is positive ____________ from left to right.

A line whose slope is negative ____________ from left to right.

A line with zero slope is __________________.

A line with undefined slope is ________________.

II. The Point-Slope Form of the Equation of a Line (Pages 5–6)

> ***What you should learn***
> How to write linear equations given points on lines and their slopes

The **point-slope form** of the equation of a line is

____________________________________ .

This form of equation is best used to find the equation of a line when ____

____.

The **two-point form** of the equation of a line is

______________________________.

The two-point form of equation is best used to find the equation of a line when ______________________________

______________________________.

Example 1: Find an equation of the line having slope – 2 that passes through the point (1, 5).

The approximation method used to estimate a point between two given points is called ____________________. The approximation method used to estimate a point that does not lie between two given points is called ____________________.

A **linear function** has the form ____________________. Its graph is a ______________ that has slope __________ and a y-intercept at

______________.

III. Sketching Graphs of Lines (Pages 7–8)

What you should learn
How to use slope-intercept forms of linear equations to sketch lines

The **slope-intercept form** of the equation of a line is ________________, where m is the ________________ and the y-intercept is (_____, _____).

Example 2: Determine the slope and y-intercept of the linear equation $2x - y = 4$.

The equation of a **horizontal line** is __________. The slope of a horizontal line is ______. The y-coordinate of every point on the graph of a horizontal line is ______________.
The equation of a **vertical line** is __________. The slope of a vertical line is ______________. The x-coordinate of every point on the graph of a vertical line is ______________.

The **general form** of the equation of a line is

______________________________.

Every line has an equation that can be written in ____________ ______________.

When a graphing utility is used to sketch a straight line, the graph of the line may not visually appear to have the slope indicated by its equation because ________________________ ____.

Example 3: Use a graphing utility to graph the linear equation $2x - y = 4$ using (a) a standard viewing window, and (b) a square window.

IV. Parallel and Perpendicular Lines (Pages 9–10)

> ***What you should learn***
> How to use slope to identify parallel and perpendicular lines

The relationship between the slopes of two lines that are parallel is ____.

The relationship between the slopes of two lines that are perpendicular is ____________________________________ ____.

A line that is parallel to a line whose slope is 2 has slope ______.
A line that is perpendicular to a line whose slope is 2 has slope _________.

What must be done to make the graphs of two perpendicular lines appear to intersect at right angles when they are graphed using a graphing utility?

Example 4: Use a graphing utility to graph the perpendicular lines $y = 2x - 3$ and $y = -0.5x + 5$ using (a) a standard viewing window, and (b) a square window.

Additional notes

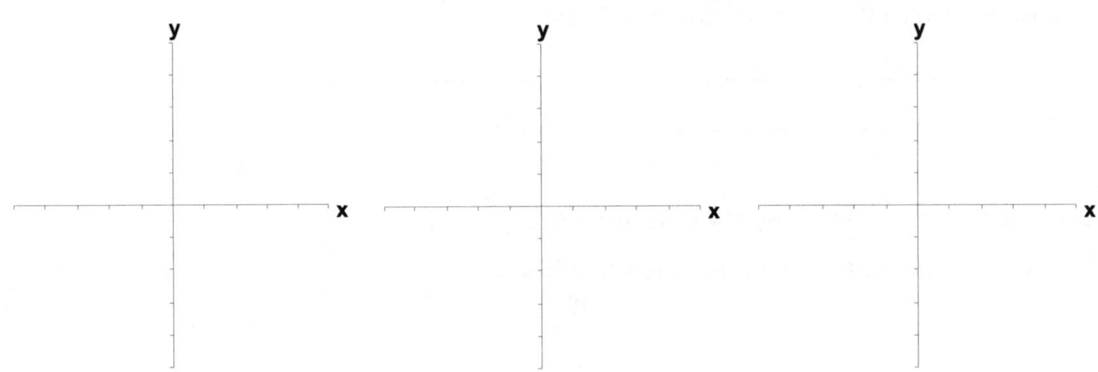

Homework Assignment

Page(s)

Exercises

Name ____ Date ___________

Section 1.2 Functions

Objective: In this lesson you learned how to evaluate functions and find their domains.

Important Vocabulary	Define each term or concept.
Function	
Domain	
Range	
Independent variable	
Dependent variable	

I. Introduction to Functions (Pages 16–17)

What you should learn
How to decide whether a relation between two variables represents a function

A rule of correspondence that matches quantities from one set with items from a different set is a ________________.

In functions that can be represented by ordered pairs, the first coordinate in each ordered pair is the ______________ and the second coordinate is the ________________.

Some characteristics of a function from Set A to Set B are

1)

2)

3)

4)

To determine whether or not a relation is a function, _______
____
__________________________.

If any input value of a relation is matched with two or more output values, ____.

Some common ways to represent functions are

1)

2)

3)

4)

Example 1: Decide whether the table represents y as a function of x.

x	-3	-1	0	2	4
y	5	-12	5	3	14

II. Function Notation (Pages 18–19)

> ***What you should learn***
> How to use function notation and evaluate functions

The symbol ______________ is **function notation** for the value of f at x or simply f of x. The symbol $f(x)$ corresponds to the ________________ for a given x.

Keep in mind that ____________ is the name of the function, whereas ___________ is the output value of the function at the input value x.

In function notation, the _____________ is the independent variable and the ____________ is the dependent variable.

Example 2: If $f(w) = 4w^3 - 5w^2 - 7w + 13$, describe how to find $f(-2)$.

A piecewise-defined function is ___________________________

____.

III. The Domain of a Function (Page 20–21)

> ***What you should learn***
> How to find the domains of functions

The **implied domain** of a function defined by an algebraic expression is ______________________________

______________________________.

In general, the domain of a function excludes values that

____

______________________________.

For example, the implied domain of the function $f(x) = \sqrt{5x-8}$ is ______________________________

______________________________.

IV. Applications of Functions (Page 22)

> ***What you should learn***
> How to use functions to model and solve real-life problems

Example 3: The price P (in dollars) of a child's handmade sweater is given by the function $P(s) = 3s + 15$, where s represents the size (size 1, size 2, etc.) of the sweater. Use this function to find the price of a child's size 5 handmade sweater.

V. Difference Quotients (Page 23)

> ***What you should learn***
> How to evaluate difference quotients

A **difference quotient** is defined as

______________________________.

Describe a real-life situation which can be represented by a function.

Additional notes

Additional notes

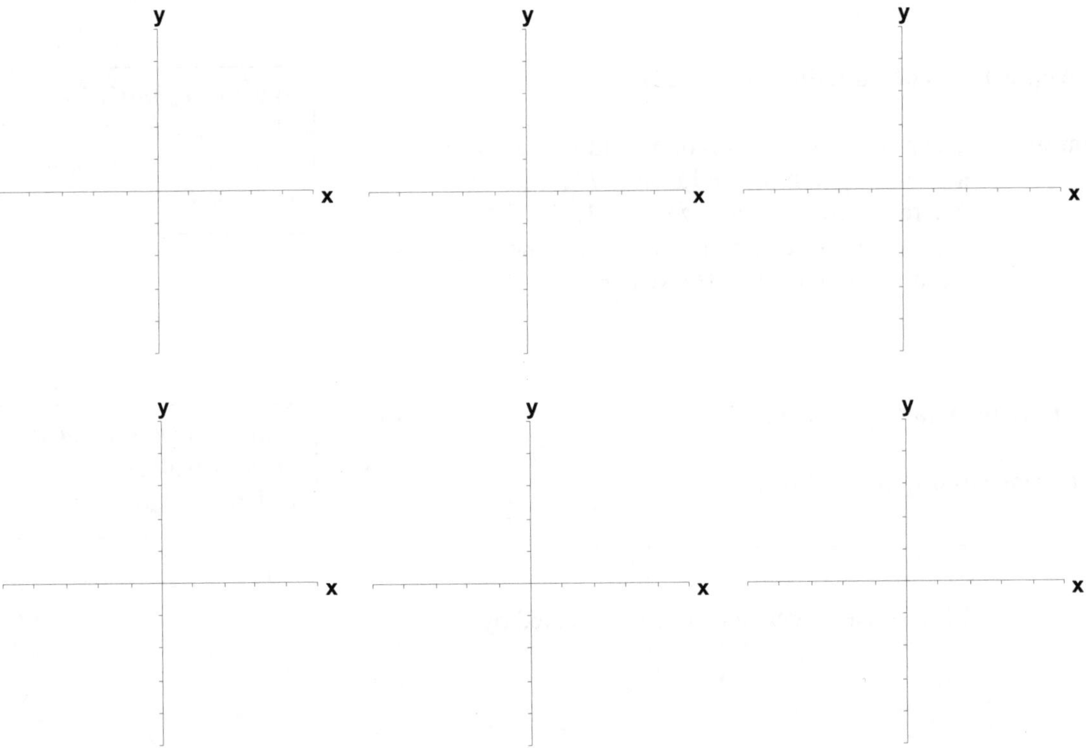

Homework Assignment

Page(s)

Exercises

Name ____ Date ____________

Section 1.3 Graphs of Functions

Objective: In this lesson you learned how to analyze the graphs of functions.

Important Vocabulary Define each term or concept.

Graph of a function

Greatest integer function

Step function

Even function

Odd function

I. The Graph of a Function (Pages 29–30)

What you should learn
How to find the domains and ranges of functions and how to use the Vertical Line Test for functions

Explain the use of open or closed dots in the graphs of functions.

How do you find the domain of a function from its graph?

How do you find the range of a function from its graph?

State the **Vertical Line Test** for functions.

Example 1: Decide whether each graph represents y as a function of x.

(a)

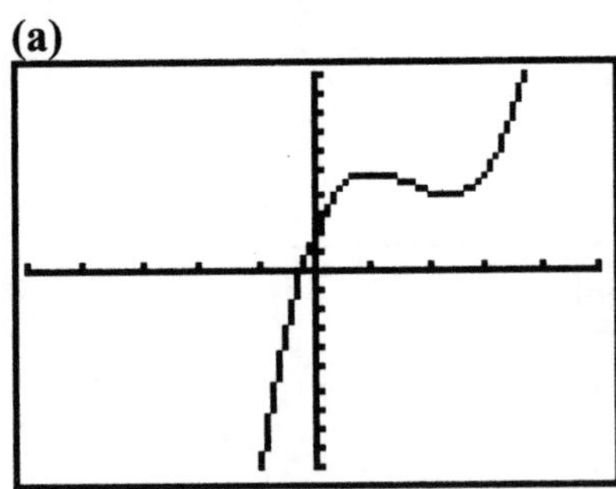

(b)

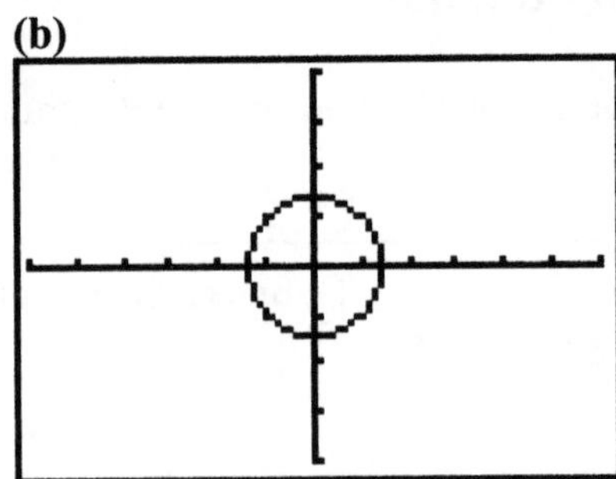

II. Increasing and Decreasing Functions (Page 31)

What you should learn
How to determine intervals on which functions are increasing, decreasing, or constant

A function f is **increasing** on an interval if, for any x_1 and x_2 in the interval, ______________________________ .

A function f is **decreasing** on an interval if, for any x_1 and x_2 in the interval, ______________________________ .

A function f is **constant** on an interval if, for any x_1 and x_2 in the interval, ________________________ .

Given a graph of a function, to find an interval on which the function is increasing ______________________________ ____.

Given a graph of a function, to find an interval on which the function is decreasing ______________________________ ____.

Given a graph of a function, to find an interval on which the function is constant ______________________________ ____.

III. Relative Minimum and Maximum Values (Pages 32–33)

What you should learn
How to determine relative maximum and relative minimum values of functions

A function value $f(a)$ is called a **relative minimum** of f if

____

____.

A function value $f(a)$ is called a **relative maximum** of f if

____

____

The point at which a function changes from increasing to decreasing is a relative ______________. The point at which a function changes from decreasing to increasing is a relative ______________.

To approximate the relative minimum or maximum of a function using a graphing utility, ______________________________

____

____

______________________________.

Example 2: Suppose a function C represents the annual number of cases (in millions) of chicken pox reported for the year x in the United States from 1960 through 2000. Interpret the meaning of the function's minimum at (1998, 3).

IV. Step Functions and Piecewise-Defined Functions (Page 34)

What you should learn
How to identify and graph step functions and other piecewise-defined functions

Describe the graph of the greatest integer function.

Example 3: Let $f(x) = [\![x]\!]$, the greatest integer function. Find $f(3.74)$.

To sketch the graph of a piecewise-defined function, ________

____

____.

V. Even and Odd Functions (Pages 35–36)

What you should learn
How to identify even and odd functions

A graph is symmetric with respect to the y-axis if, whenever (x, y) is on the graph, ____________ is also on the graph. A graph is symmetric with respect to the x-axis if, whenever (x, y) is on the graph, ____________ is also on the graph. A graph is symmetric with respect to the origin if, whenever (x, y) is on the graph, ______________ is also on the graph. A graph that is symmetric with respect to the x-axis is

____

A function f is **even** if, for each x in the domain of f,
$f(-x) =$ ________________________.

A function f is **even** if, for each x in the domain of f,
$f(-x) =$ ________________________.

Additional notes

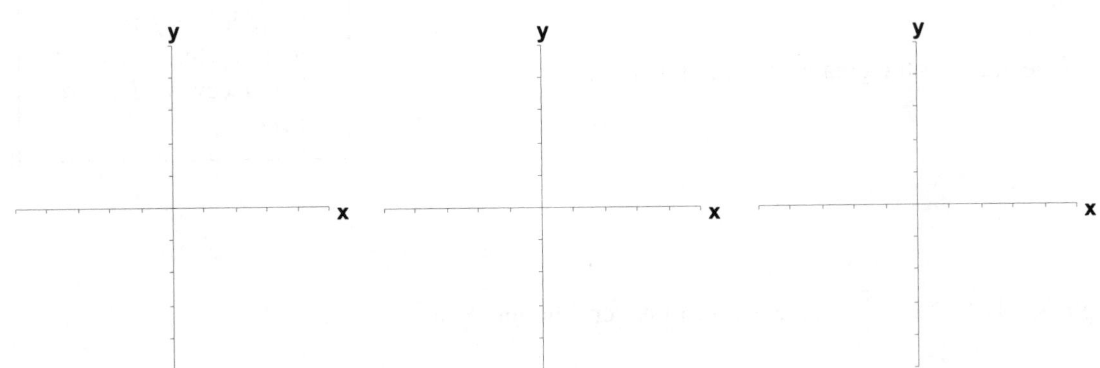

Homework Assignment

Page(s)

Exercises

Name ____ Date ____________

Section 1.4 Shifting, Reflecting, and Stretching Graphs

Objective: In this lesson you learned how to identify and graph shifts, reflections, and nonrigid transformations of functions.

Important Vocabulary Define each term or concept.

Vertical shift

Horizontal shift

Rigid transformations

Nonrigid transformations

I. Summary of Graphs of Parent Functions (Page 41)

What you should learn
How to recognize graphs of parent functions

Sketch an example of each of the six most commonly used functions in algebra.

Linear Function

Absolute Value Function

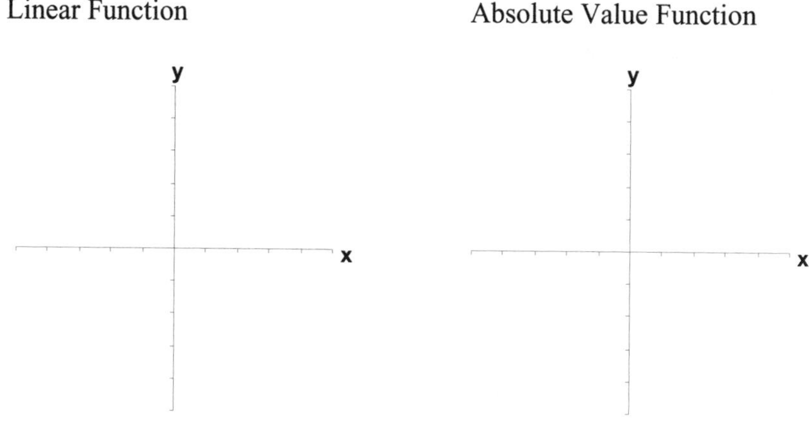

Square Root Function

Quadratic Function

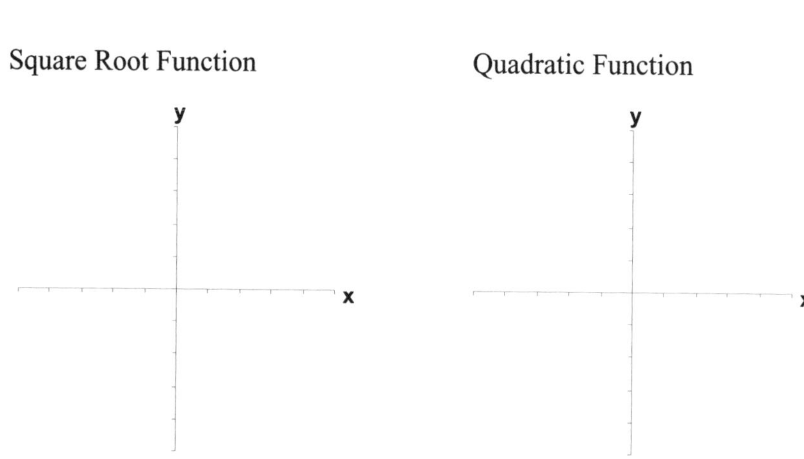

Cubic Function

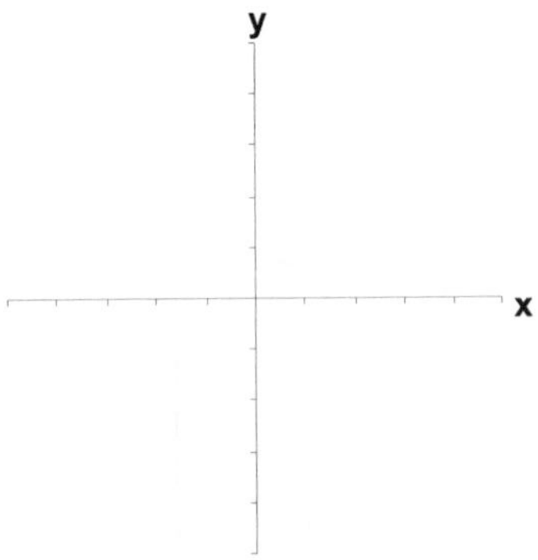

Rational Function

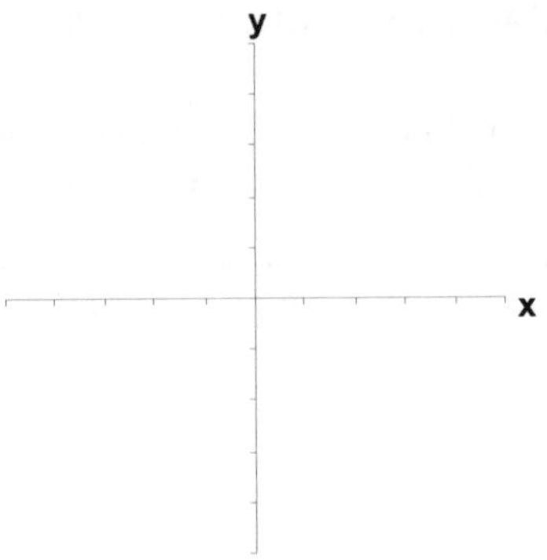

II. Vertical and Horizontal Shifts (Pages 42–43)

> ***What you should learn***
> How to use vertical and horizontal shifts and reflections to graph functions

Let c be a positive real number. Complete the following representations of shifts in the graph of $y = f(x)$:

1) Vertical shift c units upward: ________________________

2) Vertical shift c units downward: ______________________

3) Horizontal shift c units to the right: __________________

4) Horizontal shift c units to the left: ___________________

Example 1: Let $f(x) = |x|$. Write the equation for the function resulting from a vertical shift of 3 units downward and a horizontal shift of 2 units to the right of the graph of $f(x)$.

III. Reflecting Graphs (Pages 44–45)

A **reflection** in the x-axis is a type of transformation of the graph of $y = f(x)$ represented by $h(x) =$ _____________. A **reflection** in the y-axis is a type of transformation of the graph of $y = f(x)$ represented by $h(x) =$ _____________.

Example 2: Let $f(x) = |x|$. Describe the graph of $g(x) = -|x|$ in terms of f.

IV. Nonrigid Transformations (Page 46)

What you should learn
How to use nonrigid transformations to graph functions

Name three types of rigid transformations:

1)

2)

3)

Rigid transformations change only the ______________ of the graph in the coordinate plane.

Name four types of nonrigid transformations:

1)

2)

3)

4)

A nonrigid transformation $y = cf(x)$ of the graph of $y = f(x)$ is a ____________________ if $c > 1$ or a ________________ if $0 < c < 1$. A nonrigid transformation $y = f(cx)$ of the graph of $y = f(x)$ is a __________________________ if $c > 1$ or a __________________________ if $0 < c < 1$.

Additional notes

y

x

y

x

y

x

Additional notes

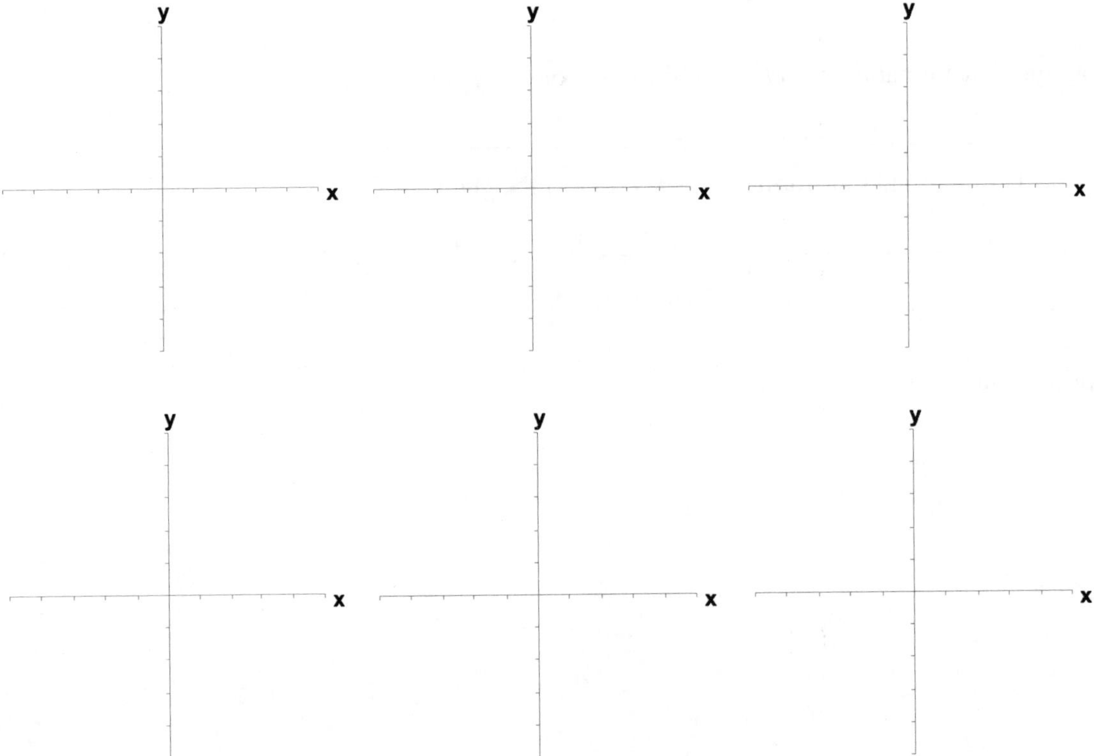

Homework Assignment

Page(s)

Exercises

Name ____ Date ____________

Section 1.5 Combinations of Functions

Objective: In this lesson you learned how to find arithmetic combinations and compositions of functions.

I. Arithmetic Combinations of Functions (Pages 50–51)

What you should learn
How to add, subtract, multiply, and divide functions

Just as two real numbers can be combined by the operations of addition, subtraction, multiplication, and division to form other real numbers, two functions *f* and *g* can be combined to create new functions such as the ______________________________ ________________ of *f* and *g* to create new functions.

The domain of an arithmetic combination of functions *f* and *g* consists of ____
____
____.

Let *f* and *g* be two functions with overlapping domains. Complete the following arithmetic combinations of *f* and *g* for all *x* common to both domains:

1) Sum: $(f+g)(x) =$ ____________________

2) Difference: $(f-g)(x) =$ ____________________

3) Product: $(fg)(x) =$ ____________________

4) Quotient: $\left(\frac{f}{g}\right)(x) =$ ____________________

To use a graphing utility to graph the sum of two functions,
____
____

Example 1: Let $f(x) = 7x - 5$ and $g(x) = 3 - 2x$. Find $(f-g)(4)$.

II. Compositions of Functions (Pages 52–54)

> ***What you should learn***
> How to find compositions of one function with another function

The **composition** of the function f with the function g is

$(f \circ g)(x) =$ ________________________.

For the composition of the function f with g, the domain of

$f \circ g$ is ____

____.

Example 2: Let $f(x) = 3x + 4$ and let $g(x) = 2x^2 - 1$. Find (a) $(f \circ g)(x)$ and (b) $(g \circ f)(x)$.

III. Applications of Combinations of Functions (Page 55)

> ***What you should learn***
> How to use combinations of functions to model and solve real-life problems

The function $f(x) = 0.06x$ represents the sales tax owed on a purchase with a price tag of x dollars and the function $g(x) = 0.75x$ represents the sale price of an item with a price tag of x dollars during a 25% off sale. Using one of the combinations of functions discussed in this section, write the function that represents the sales tax owed on an item with a price tag of x dollars during a 25% off sale.

Additional notes

Homework Assignment

Page(s)

Exercises

Name ____ Date ____________

Section 1.6 Inverse Functions

Objective: In this lesson you learned how to find inverse functions graphically and algebraically.

Important Vocabulary Define each term or concept.

Inverse function

One-to-one

Horizontal Line Test

I. Inverse Functions (Pages 60–62)

> ***What you should learn***
> How to find inverse functions informally and verify that two functions are inverse functions of each other

For a function f that is defined by a set of ordered pairs, to form the inverse function of f, ______________________________

____.

For a function f and its inverse f^{-1}, the domain of f is equal to ______________________, and the range of f is equal to ______________________.

To verify that two functions, f and g, are inverses of each other,

____

____

____.

Example 1: Verify that the functions $f(x) = 2x - 3$ and $g(x) = \dfrac{x+3}{2}$ are inverses of each other.

II. The Graph of an Inverse Function (Page 63)

> ***What you should learn***
> How to use graphs of functions to decide whether functions have inverse functions

If the point (a, b) lies on the graph of f, then the point (____________) lies on the graph of f^{-1} and vice versa. The graph of f^{-1} is a reflection of the graph of f in the line

______________.

III. The Existence of an Inverse Function (Page 64)

What you should learn
How to determine whether functions are one-to-one

If a function is **one-to-one,** that means ____________________
____
____.

A function f has an inverse f^{-1} if and only if ________________
____________________.

To tell whether a function is one-to-one from its graph, ______
____
____

Example 2: Does the graph of the function at the right have an inverse function? Explain.

IV. Finding Inverse Functions Algebraically (Pages 65–66)

What you should learn
How to find inverse functions algebraically

To find the inverse of a function f algebraically,

1)

2)

3)

4)

5)

Example 3: Find the inverse (if it exists) of $f(x) = 4x - 5$.

Homework Assignment

Page(s)

Exercises

Name ____ Date __________

Section 1.7 Linear Models and Scatter Plots

Objective: In this lesson you learned how to use scatter plots and a graphing utility to find linear models for data.

Important Vocabulary	Define each term or concept.
Fitting a line to data	

I. Scatter Plots and Correlation (Pages 71–72)

What you should learn
How to construct scatter plots and interpret correlation

Many real-life situations involve finding relationships between two variables. If data are collected and written as a set of ordered pairs, the graph of such a set is called a ________________.

For a collection of ordered pairs of the form (x, y), if y tends to increase as x increases, then the collection is said to have a(n) ____________________________. If y tends to decrease as x increases, then the collection is said to have a(n) __________ ________________.

II. Fitting a Line to Data (Pages 73–75)

What you should learn
How to use scatter plots and a graphing utility to find linear models for data

Describe how to fit a line to data represented in a scatter plot.

To measure how well a linear model fits the data used to find the model, ____ ____________________________.

The correlation coefficient r of a set of data varies between

________ and ________. The closer $|r|$ is to 1, the better ______

____.

Example 1: The numbers of U.S. Navy personnel p in thousands on active duty for the years 2002 through 2006 are shown in the table. Use the regression capabilities of a graphing utility to find a linear model for the data. Let t represent the year with $t = 2$ corresponding to 2002.

Year	2002	2003	2004	2005	2006
p	383	381	376	364	353

(Source: U.S. Department of Defense)

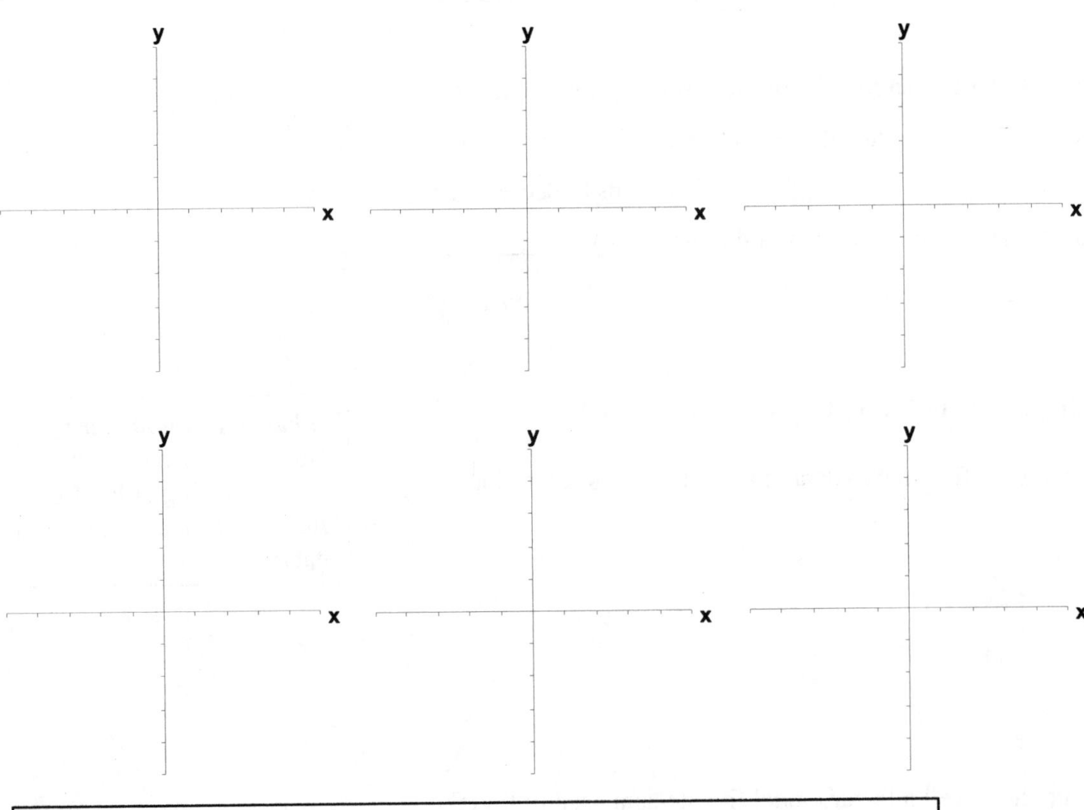

Homework Assignment

Page(s)

Exercises

Name ____ Date __________

Chapter 2 Polynomial and Rational Functions

Section 2.1 Quadratic Functions

Objective: In this lesson you learned how to sketch and analyze graphs of quadratic functions.

Important Vocabulary	Define each term or concept.
Constant function	
Linear function	
Quadratic function	
Axis of symmetry	
Vertex	

I. The Graph of a Quadratic Function (Pages 90–92)

What you should learn
How to analyze graphs of quadratic functions

Let n be a nonnegative integer and let $a_n, a_{n-1}, \ldots, a_2, a_1, a_0$ be real numbers with $a_n \neq 0$. A **polynomial function of *x* with degree *n*** is

____

A quadratic function is a polynomial function of ____________ degree. The graph of a quadratic function is a special "U"-shaped curve called a(n) __________________.

If the leading coefficient of a quadratic function is positive, the graph of the function opens ____________ and the vertex of the parabola is the ______________ point on the graph. If the leading coefficient of a quadratic function is negative, the graph of the function opens _______________ and the vertex of the parabola is the _______________ point on the graph.

II. The Standard Form of a Quadratic Function (Pages 93–94)

> ***What you should learn***
> How to write quadratic functions in standard form and use the results to sketch graphs of functions

The **standard form of a quadratic function** is

____ .

For a quadratic function in standard form, the axis of the associated parabola is ___________ and the vertex is

______________.

To write a quadratic function in standard form, __________

____.

To find the x-intercepts of the graph of $f(x) = ax^2 + bx + c$,

____.

Example 1: Sketch the graph of $f(x) = x^2 + 2x - 8$ and identify the vertex, axis, and x-intercepts of the parabola.

y

x

III. Finding Minimum and Maximum Values (Page 95)

> ***What you should learn***
> How to find minimum and maximum values of quadratic functions in real-life applications

For a quadratic function in the form $f(x) = ax^2 + bx + c$, when $a > 0$, f has a minimum that occurs at ____________________.
When $a < 0$, f has a maximum that occurs at ________________.
To find the minimum or maximum value, ___________________

____.

Example 2: Find the minimum value of the function $f(x) = 3x^2 - 11x + 16$. At what value of x does this minimum occur?

Homework Assignment

Page(s)

Exercises

Name ______________________________ Date __________

Section 2.2 Polynomial Functions of Higher Degree

Objective: In this lesson you learned how to sketch and analyze graphs of polynomial functions.

Important Vocabulary Define each term or concept.

Continuous

Extrema

Relative minimum

Relative maximum

Repeated zero

Multiplicity

Intermediate Value Theorem

I. Graphs of Polynomial Functions (Pages 100–101)

What you should learn
How to use transformations to sketch graphs of polynomial functions

Name two basic features of the graphs of polynomial functions.

1)
2)

Will the graph of $g(x) = x^7$ look more like the graph of $f(x) = x^2$ or the graph of $f(x) = x^3$? Explain.

II. The Leading Coefficient Test (Pages 102–103)

What you should learn
How to use the Leading Coefficient Test to determine the end behavior of graphs of polynomial functions

State the **Leading Coefficient Test.**

1.
   a.
   b.
2.
   a.
   b.

Example 1: Describe the left and right behavior of the graph of $f(x) = 1 - 3x^2 - 4x^6$.

III. Zeros of Polynomial Functions (Pages 104–107)

What you should learn
How to find and use zeros of polynomial functions as sketching aids

Let f be a polynomial function of degree n. The function f has at most __________ real zeros. The graph of f has at most __________ relative extrema.

Let f be a polynomial function and let a be a real number. List four equivalent statements about the real zeros of f.
1)
2)
3)
4)

If a polynomial function f has a repeated zero $x = 3$ with multiplicity 4, the graph of f ____________ the x-axis at $x =$ ______. If f has a repeated zero $x = 4$ with multiplicity 3, the graph of f ____________ the x-axis at $x =$ ______.

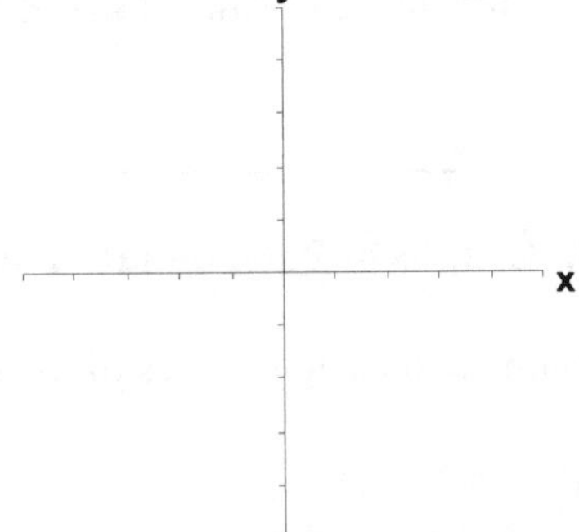

Example 2: Sketch the graph of $f(x) = \frac{1}{4}x^4 - 2x^2 + 3$.

IV. The Intermediate Value Theorem (Page 108)

What you should learn
How to use the Intermediate Value Theorem to help locate zeros of polynomial functions

Interpret the meaning of the Intermediate Value Theorem.

Describe how the Intermediate Value Theorem can help in locating the real zeros of a polynomial function f.

Homework Assignment

Page(s)

Exercises

Name ______________________________ Date __________

Section 2.3 Real Zeros of Polynomial Functions

Objective: In this lesson you learned how to use long division and synthetic division to divide polynomials by other polynomials and how to find the rational and real zeros of polynomial functions.

Important Vocabulary Define each term or concept.

Long division of polynomials

Division Algorithm

Synthetic division

Remainder Theorem

Factor Theorem

Upper bound

Lower bound

I. Long Division of Polynomials (Pages 113–115)

What you should learn
How to use long division to divide polynomials by other polynomials

When dividing a polynomial $f(x)$ by another polynomial $d(x)$, if the remainder $r(x) = 0$, $d(x)$ ____________ into $f(x)$.

The rational expression $f(x)/d(x)$ is **improper** if ____________
____________.

The rational expression $r(x)/d(x)$ is **proper** if ____________
____________.

Before applying the Division Algorithm, you should ____________
____
____
____________.

Example 1: Divide $3x^3 + 4x - 2$ by $x^2 + 2x + 1$.

II. Synthetic Division (Page 116)

What you should learn
How to use synthetic division to divide polynomials by binomials of the form $(x - k)$

Can synthetic division be used to divide a polynomial by $x^2 - 5$? Explain.

Can synthetic division be used to divide a polynomial by $x + 4$? Explain.

Example 2: Fill in the following synthetic division array to divide $2x^4 + 5x^2 - 3$ by $x - 5$. Then carry out the synthetic division and indicate which entry represents the remainder.

III. The Remainder and Factor Theorems (Pages 117–118)

What you should learn
How to use the Remainder and Factor Theorems

To use the Remainder Theorem to evaluate a polynomial function $f(x)$ at $x = k$, ______________________________

____.

Example 3: Use the Remainder Theorem to evaluate the function $f(x) = 2x^4 + 5x^2 - 3$ at $x = 5$.

To use the Factor Theorem to show that $(x - k)$ is a factor of a polynomial function $f(x)$, ______________________________

____

____

____.

List three facts about the remainder r, obtained in the synthetic division of $f(x)$ by $x - k$:
1)

2)

3)

IV. The Rational Zero Test (Pages 119–120)

> ***What you should learn***
> How to use the Rational Zero Test to determine possible rational zeros of polynomial functions

Describe the purpose of the Rational Zero Test.

State the **Rational Zero Test.**

Describe how to use the Rational Zero Test.

Example 4: List the possible rational zeros of the polynomial function $f(x) = 3x^5 + x^4 + 4x^3 - 2x^2 + 8x - 5$.

List some strategies that can be used to shorten the search for actual zeros among a list of possible rational zeros.

V. Other Tests for Zeros of Polynomials (Pages 121–123)

State the Upper and Lower Bound Rules.

1.

2.

Explain how the Upper and Lower Bound Rules can be useful in the search for the real zeros of a polynomial function.

What you should learn
How to use Descartes's Rule of Signs and the Upper and Lower Bound Rules to find zeros of polynomials

Additional notes

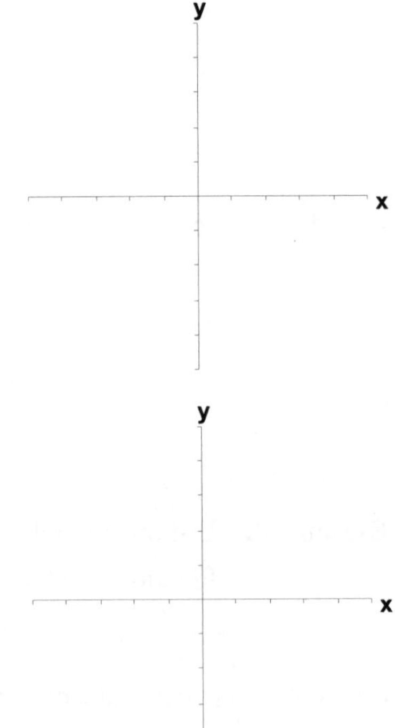

Homework Assignment

Page(s)

Exercises

Name ____ Date ____________

Section 2.4 Complex Numbers

Objective: In this lesson you learned how to perform operations with complex numbers.

Important Vocabulary	Define each term or concept.
Complex number	
Complex conjugates	

I. The Imaginary Unit i (Page 128)

What you should learn
How to use the imaginary unit i to write complex numbers

Mathematicians created an expanded system of numbers using the **imaginary unit** i, defined as i = ___________, because

____

By definition, i^2 = ____________.

For the complex number $a + bi$, if $b = 0$, the number $a + bi = a$ is a(n) __________________. If $b \neq 0$, the number $a + bi$ is a(n) __________________________. If $a = 0$, the number $a + bi = b$, where $b \neq 0$, is called a(n) __________________________.

The set of complex numbers consists of the set of ________ ___________ and the set of __________________________.

Two complex numbers $a + bi$ and $c + di$, written in standard form, are equal to each other if __________________________ .

II. Operations with Complex Numbers (Pages 129–130)

What you should learn
How to add, subtract, and multiply complex numbers

To add two complex numbers, ____________________________

____.

To subtract two complex numbers, _______________________

____.

The additive identity in the complex number system is ___________.

The additive inverse of the complex number $a + bi$ is

______________________.

Example 1: Perform the operations:
$(5-6i)-(3-2i)+4i$

To multiply two complex numbers $a + bi$ and $c + di$, ________

____

____

____.

Example 2: Multiply: $(5-6i)(3-2i)$

III. Complex Conjugates (Page 131)

> ***What you should learn***
> How to use complex conjugates to write the quotient of two complex numbers in standard form

The product of a pair of complex conjugates is a(n)

____________________ number.

To find the quotient of the complex numbers $a + bi$ and $c + di$, where c and d are not both zero, ________________________

____.

Example 3: Divide $(1 + i)$ by $(2 - i)$. Write the result in standard form.

IV. Complex Solutions of Quadratic Equations (Page 132)

> ***What you should learn***
> How to find complex solutions of quadratic equations

When using the Quadratic Formula to solve a quadratic equation, you may obtain a result such as $\sqrt{-7}$, which is not a__________

______________. By factoring out $i = \sqrt{-1}$, you can write this number in ________________________________.

If a is a positive number, then the **principal square root** of the negative number $-a$ is defined as ______________.

> **Homework Assignment**
> Page(s)
> Exercises

Name ______________________________ Date __________

Section 2.5 The Fundamental Theorem of Algebra

Objective: In this lesson you learned how to determine the numbers of zeros of polynomial functions and find them.

Important Vocabulary Define each term or concept.

Fundamental Theorem of Algebra

Linear Factorization Theorem

Conjugates

I. The Fundamental Theorem of Algebra (Page 135)

What you should learn
How to use the Fundamental Theorem of Algebra to determine the number of zeros of a polynomial function

In the complex number system, every nth-degree polynomial function has ________________ zeros.

Example 1: How many zeros does the polynomial function $f(x) = 5 - 2x^2 + x^3 - 12x^5$ have?

An nth-degree polynomial can be factored into ________________ linear factors.

II. Finding Zeros of a Polynomial Function (Page 136)

What you should learn
How to find all zeros of polynomial functions, including complex zeros

Remember that the n zeros of a polynomial function can be real or complex, and they may be ____________________.

Example 2: List all of the zeros of the polynomial function $f(x) = x^3 - 2x^2 + 36x - 72$.

III. Conjugate Pairs (Page 137)

What you should learn
How to find conjugate pairs of complex zeros

Let $f(x)$ be a polynomial function that has real coefficients. If $a + bi$, where $b \neq 0$, is a zero of the function, then we know that ______________ is also a zero of the function.

IV. Factoring a Polynomial (Pages 138–139)

What you should learn
How to find zeros of polynomials by factoring

To write a polynomial of degree $n > 0$ with real coefficients as a product without complex factors, write the polynomial as ______
____
____.

A quadratic factor with no real zeros is said to be
____.

Example 3: Write the polynomial $f(x) = x^4 + 5x^2 - 36$
(a) as the product of linear factors and quadratic factors that are irreducible over the reals, and
(b) in completely factored form.

Explain why a graph cannot be used to locate complex zeros.

Additional notes

y

x

Homework Assignment

Page(s)

Exercises

Name ____ Date __________

Section 2.6 Rational Functions and Asymptotes

Objective: In this lesson you learned how to determine the domains and find asymptotes of rational functions.

Important Vocabulary	Define each term or concept.
Rational function	
Vertical asymptote	
Horizontal asymptote	

I. Introduction to Rational Functions (Page 142)

> ***What you should learn***
> How to find the domains of rational functions

The domain of a rational function of x includes all real numbers except ______________________________________.

To find the domain of a rational function of x, ______________
____
____
____________________________________.

Example 1: Find the domain of the function $f(x) = \dfrac{1}{x^2 - 9}$.

II. Vertical and Horizontal Asymptotes (Pages 143–145)

> ***What you should learn***
> How to find vertical and horizontal asymptotes of graphs of rational functions

The notation "$f(x) \to 5$ as $x \to \infty$" means ______________
____________________________________.

Let f be the rational function given by

$$f(x) = \frac{N(x)}{D(x)} = \frac{a_n x^n + a_{n-1}x^{n-1} + \cdots + a_1 x + a_0}{b_m x^m + b_{m-1}x^{m-1} + \cdots + b_1 x + b_0}$$

where $N(x)$ and $D(x)$ have no common factors.

1) The graph of f has vertical asymptotes at ____________
__________________.

2) The graph of f has at most one horizontal asymptote determined by ________________________________

____________.

a) If $n < m$, ____________________________________

____.

b) If $n = m$, ______________________________________

____________________ .

c) If $n > m$, the graph of f has ________________

______________.

Example 2: Find the asymptotes of the function

$$f(x) = \frac{2x-1}{x^2 - x - 6}.$$

III. Application of Rational Functions (Page 146)

> ***What you should learn***
> How to use rational functions to model and solve real-life problems

Give an example of asymptotic behavior that occurs in real life.

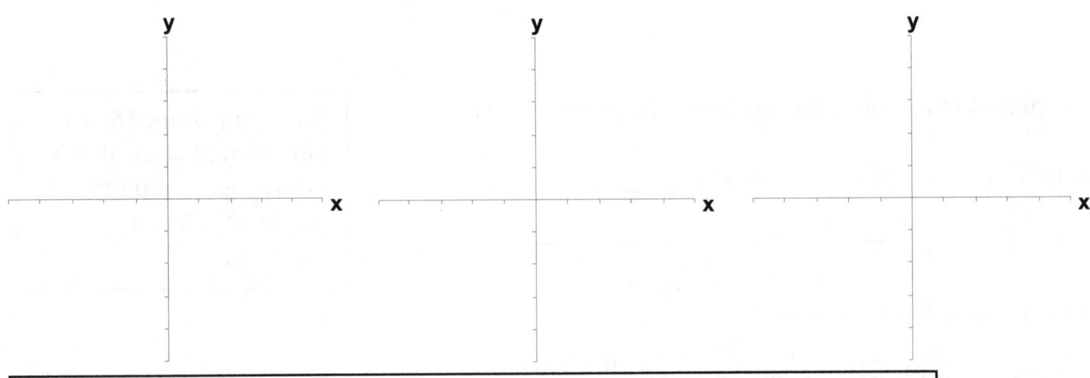

Homework Assignment

Page(s)

Exercises

Name ______________________________ Date __________

Section 2.7 Graphs of Rational Functions

Objective: In this lesson you learned how to sketch graphs of rational functions.

Important Vocabulary Define each term or concept.

Slant (or oblique) asymptote

I. The Graph of a Rational Function (Pages 151–154)

What you should learn
How to analyze and sketch graphs of rational functions

List the guidelines for sketching the graph of the rational function $f(x) = N(x)/D(x)$, where $N(x)$ and $D(x)$ are polynomials.

Example 1: Sketch the graph of $f(x) = \dfrac{3x}{x+4}$.

y

x

II. Slant Asymptotes (Page 155)

> ***What you should learn***
> How to sketch graphs of rational functions that have slant asymptotes

Describe how to find the equation of a slant asymptote.

Example 2: Decide whether each of the following rational functions has a slant asymptote. If so, find the equation of the slant asymptote.

(a) $f(x) = \dfrac{x^3 - 1}{x^2 + 3x + 5}$ (b) $f(x) = \dfrac{3x^3 + 2}{2x - 5}$

III. Applications of Graphs of Rational Functions (Page 156)

> ***What you should learn***
> How to use graphs of rational functions to model and solve real-life problems

Describe a real-life situation in which a graph of a rational function would be helpful when solving a problem.

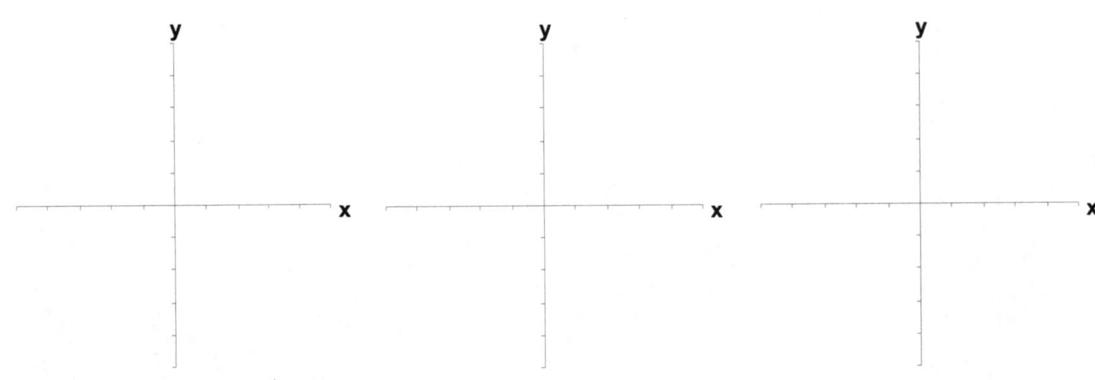

Homework Assignment

Page(s)

Exercises

Name ______________________________ Date ____________

Section 2.8 Quadratic Models

Objective: In this lesson you learned how to classify scatter plots and use a graphing utility to find quadratic models for data.

I. Classifying Scatter Plots (Page 161)

> ***What you should learn***
> How to classify scatter plots

Describe how to decide whether a set of data can be modeled by a linear model.

Describe how to decide whether a set of data can be modeled by a quadratic model.

II. Fitting a Quadratic Model to Data (Pages 162–163)

> ***What you should learn***
> How to use scatter plots and a graphing utility to find quadratic models for data

Once it has been determined that a quadratic model is appropriate for a set of data, a quadratic model can be fit to data by ______________________________

______________________________.

Example 1: Find a model that best fits the data given in the table.

x	−1	0	2	5	9	12	15
y	8.7	3.45	−5.55	−15.3	−21.3	−20.55	−15.3

III. Choosing a Model (Page 164)

What you should learn
How to choose a model that best fits a set of data

If it isn't easy to tell from a scatter plot which type of model a set of data would best be modeled by, you should ____________
____
____
____.

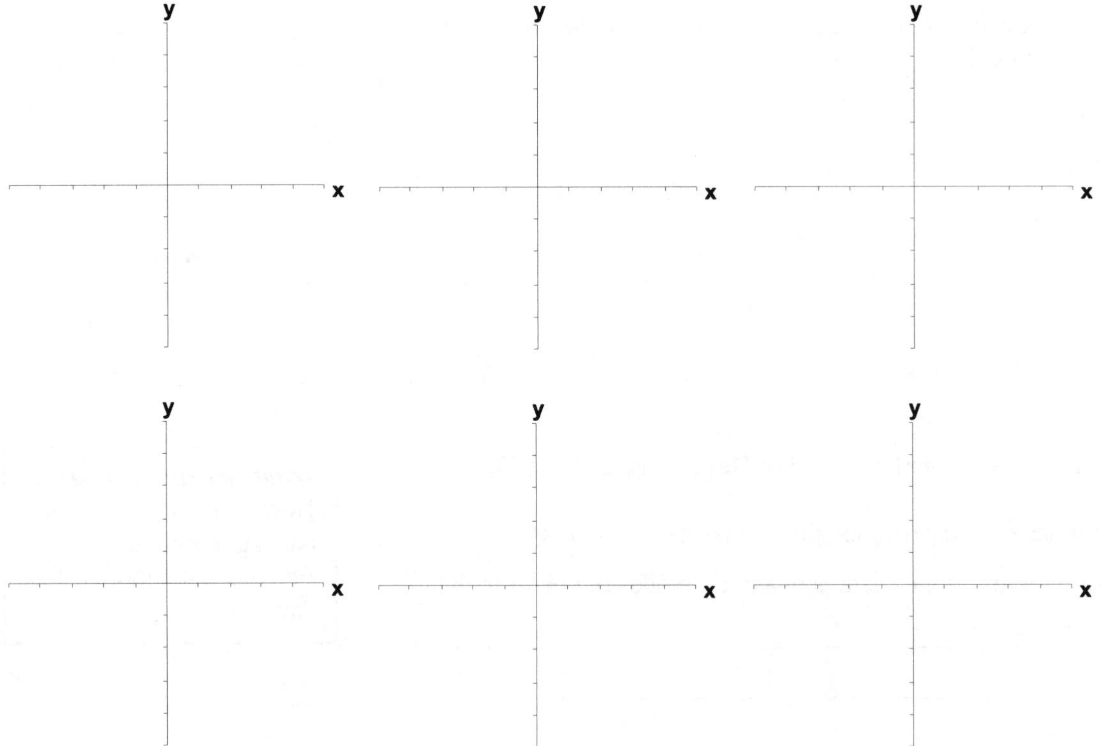

Homework Assignment

Page(s)

Exercises

Name ______________________________ Date ____________

Chapter 3 Exponential and Logarithmic Functions

Section 3.1 Exponential Functions and Their Graphs

Objective: In this lesson you learned how to recognize, evaluate, and graph exponential functions.

Important Vocabulary	Define each term or concept.
Transcendental functions	
Natural base *e*	

I. Exponential Functions (Page 180)

What you should learn
How to recognize and evaluate exponential functions with base a

Polynomial functions and rational functions are examples of ____________ functions.

The **exponential function *f* with base *a*** is denoted by ____________, where $a > 0$, $a \neq 1$, and x is any real number.

Example 1: Use a calculator to evaluate the expression $5^{3/5}$.

II. Graphs of Exponential Functions (Pages 181–183)

What you should learn
How to graph exponential functions with base a

For $a > 1$, is the graph of $f(x) = a^x$ increasing or decreasing over its domain? ____________

For $a > 1$, is the graph of $g(x) = a^{-x}$ increasing or decreasing over its domain? ____________

For the graph of $y = a^x$ or $y = a^{-x}$, $a > 1$, the domain is ____________, the range is ____________, and the intercept is ____________. Also, both graphs have ____________ as a horizontal asymptote.

Example 2: Sketch the graph of the function $f(x) = 3^{-x}$.

y

x

III. The Natural Base *e* (Pages 184–185)

> ***What you should learn***
> How to recognize, evaluate, and graph exponential functions with base *e*

The **natural exponential function** is given by the function

________________.

Example 3: Use a calculator to evaluate the expression $e^{3/5}$.

For the graph of $f(x) = e^x$, the domain is ______________,

the range is ____________, and the intercept is __________.

The number *e* can be approximated by the expression

______________ for large values of *x*.

IV. Applications (Pages 186–188)

> ***What you should learn***
> How to use exponential functions to model and solve real-life problems

After *t* years, the balance *A* in an account with principal *P* and annual interest rate *r* (in decimal form) is given by the formulas:

For *n* compoundings per year: ____________________

For continuous compounding: ______________

Example 4: Find the amount in an account after 10 years if $6000 is invested at an interest rate of 7%,
(a) compounded monthly.
(b) compounded continuously.

Homework Assignment

Page(s)

Exercises

Name ____ Date ____________

Section 3.2 Logarithmic Functions and Their Graphs

Objective: In this lesson you learned how to recognize, evaluate, and graph logarithmic functions.

Important Vocabulary	Define each term or concept.
Common logarithmic function	
Natural logarithmic function	

I. Logarithmic Functions (Pages 192–193)

> ***What you should learn***
> How to recognize and evaluate logarithmic functions with base *a*

The logarithmic function with base *a* is the ________________ ________________ of the exponential function $f(x) = a^x$.

The **logarithmic function with base *a*** is defined as ____________________ , for $x > 0$, $a > 0$, and $a \neq 1$, if and only if $x = a^y$. The notation "$\log_a x$" is read as "________________ ____________________."

The equation $x = a^y$ in exponential form is equivalent to the equation ____________________ in logarithmic form.

When evaluating logarithms, remember that a logarithm is a(n) ______________. This means that $\log_a x$ is the ______________ to which *a* must be raised to obtain ________.

Example 1: Use the definition of logarithmic function to evaluate $\log_5 125$.

Example 2: Use a calculator to evaluate $\log_{10} 300$.

Complete the following properties of logarithms:

1) $\log_a 1 =$ ____________ 2) $\log_a a =$ ____________

3) $\log_a a^x =$ ____________ and $a^{\log_a x} =$ ____________

4) If $\log_a x = \log_a y$, then ________________.

Example 3: Solve the equation $\log_7 x = 1$ for x.

II. Graphs of Logarithmic Functions (Pages 194–195)

What you should learn
How to graph logarithmic functions with base a

For $a > 1$, is the graph of $f(x) = \log_a x$ increasing or decreasing over its domain? ______________________________

For the graph of $f(x) = \log_a x$, $a > 1$, the domain is ____________________, the range is ____________________, and the intercept is ________________.

Also, the graph has ___________________________ as a vertical asymptote. The graph of $f(x) = \log_a x$ is a reflection of the graph of $f(x) = a^x$ in ______________________.

Example 4: Sketch the graph of the function $f(x) = \log_3 x$.

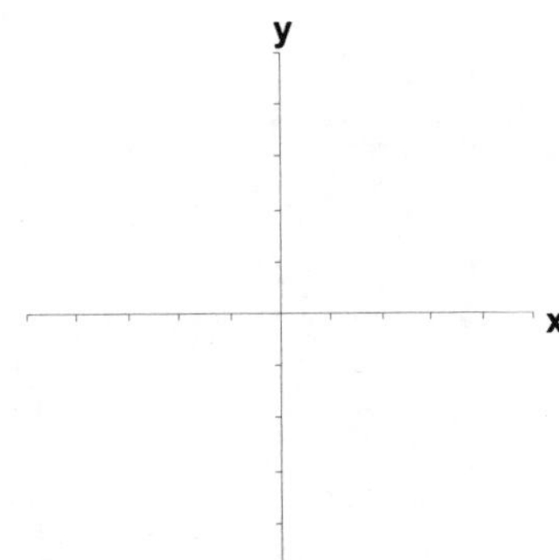

III. The Natural Logarithmic Function (Pages 196–197)

What you should learn
How to recognize, evaluate, and graph natural logarithmic functions

Complete the following properties of natural logarithms:

1) $\ln 1 =$ __________ 2) $\ln e =$ __________

3) $\ln e^x =$ __________ and $e^{\ln x} =$ __________

4) If $\ln x = \ln y$, then ______________.

Example 5: Use a calculator to evaluate $\ln 10$.

Example 6: Find the domain of the function $f(x) = \ln(x+3)$.

IV. Applications of Logarithmic Functions (Page 198)

What you should learn
How to use logarithmic functions to model and solve real-life problems

Describe a real-life situation in which logarithms are used.

Example 7: A principal *P*, invested at 6% interest and compounded continuously, increases to an amount *K* times the original principal after *t* years, where *t* is given by $t = \frac{\ln K}{0.06}$. How long will it take the original investment to double in value? To triple in value?

Additional notes

Homework Assignment

Page(s)

Exercises

Name ______________________________ Date __________

Section 3.3 Properties of Logarithms

Objective: In this lesson you learned how to rewrite logarithmic functions with different bases and how to use properties of logarithms to evaluate, rewrite, expand, or condense logarithmic expressions.

I. Change of Base (Page 203)

> ***What you should learn***
> How to rewrite logarithms with different bases

Let a, b, and x be positive real numbers such that $a \neq 1$ and $b \neq 1$. The **change-of-base formula** states that:

Explain how to use a calculator to evaluate $\log_8 20$.

II. Properties of Logarithms (Page 204)

> ***What you should learn***
> How to use properties of logarithms to evaluate or rewrite logarithmic expressions

Let a be a positive number such that $a \neq 1$; let n be a real number; and let u and v be positive real numbers. Complete the following properties of logarithms:

1. $\log_a(uv) =$ ____________________

2. $\log_a \frac{u}{v} =$ ____________________

3. $\log_a u^n =$ ____________________

III. Rewriting Logarithmic Expressions (Page 205)

> ***What you should learn***
> How to use properties of logarithms to expand or condense logarithmic expressions

To expand a logarithmic expression means to ______________

____

____

____.

Example 1: Expand the logarithmic expression $\ln \frac{xy^4}{2}$.

To condense a logarithmic expression means to ________

____

____.

Example 2: Condense the logarithmic expression $3\log x + 4\log(x-1)$.

IV. Applications of Properties of Logarithms (Page 206)

What you should learn
How to use logarithmic functions to model and solve real-life problems

One way of finding a model for a set of nonlinear data is to take the natural log of each of the x-values and y-values of the data set. If the points are graphed and fall on a straight line, then the x-values and the y-values are related by the equation:

_________________________, where m is the slope of the straight line.

Example 3: Find a natural logarithmic equation for the following data that expresses y as a function of x.

x	2.718	7.389	20.086	54.598
y	7.389	54.598	403.429	2980.958

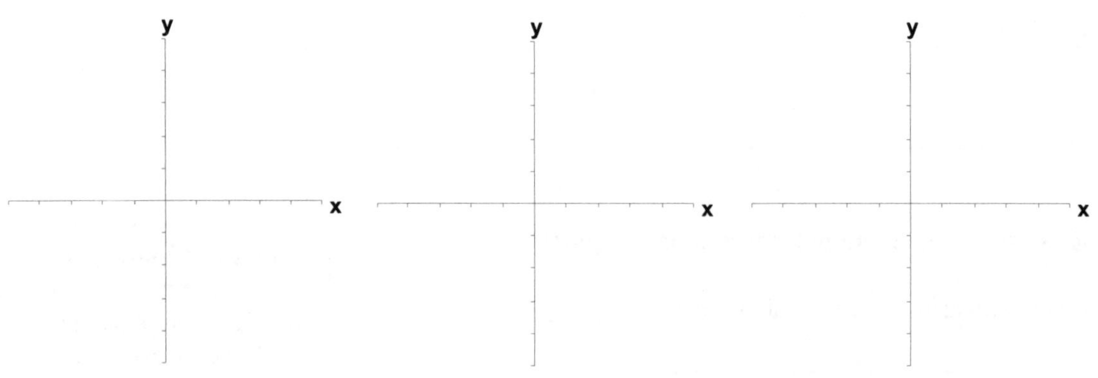

Homework Assignment

Page(s)

Exercises

Name ____ Date ___________

Section 3.4 Solving Exponential and Logarithmic Equations

Objective: In this lesson you learned how to solve exponential and logarithmic equations.

I. Introduction (Page 210)

> ***What you should learn***
> How to solve simple exponential and logarithmic equations

State the One-to-One Property for exponential equations.

State the One-to-One Property for logarithmic equations.

State the Inverse Properties for exponential equations and for logarithmic equations.

Describe some strategies for using the One-to-One Properties and the Inverse Properties to solve exponential and logarithmic equations.

-
-
-

Example 1: (a) Solve $\log_8 x = \frac{1}{3}$ for x.

(b) Solve $5^x = 0.04$ for x.

II. Solving Exponential Equations (Pages 211–212)

> ***What you should learn***
> How to solve more complicated exponential equations

Describe how to solve the exponential equation $10^x = 90$ algebraically.

Example 2: Solve $e^{x-2} - 7 = 59$ for x. Round to three decimal places.

III. Solving Logarithmic Equations (Pages 213–215)

What you should learn
How to solve more complicated logarithmic equations

Describe how to solve the logarithmic equation $\log_6(4x-7) = \log_6(8-x)$ algebraically.

Example 3: Solve $4 \ln 5x = 28$ for x. Round to three decimal places.

Describe a method that can be used to approximate the solutions of an exponential or logarithmic equation using a graphing utility.

IV. Applications of Solving Exponential and Logarithmic Equations (Page 216)

What you should learn
How to use exponential and logarithmic equations to model and solve real-life problems

Example 4: Use the formula for continuous compounding, $A = Pe^{rt}$, to find how long it will take \$1500 to triple in value if it is invested at 12% interest, compounded continuously.

Homework Assignment

Page(s)

Exercises

Name ____ Date __________

Section 3.5 Exponential and Logarithmic Models

Objective: In this lesson you learned how to use exponential growth models, exponential decay models, Gaussian models, logistic models, and logarithmic models to solve real-life problems.

Important Vocabulary Define each term or concept.

Bell-shaped curve

Logistic curve

Sigmoidal curve

I. Introduction (Page 221)

What you should learn
How to recognize the five most common types of models involving exponential or logarithmic functions

The **exponential growth model** is ____________________.

The **exponential decay model** is ____________________.

The **Gaussian model** is ____________________ .

The **logistic growth model** is ____________________ .

Logarithmic models are ________________ and

____________________.

II. Exponential Growth and Decay (Pages 222–224)

What you should learn
How to use exponential growth and decay functions to model and solve real-life problems

Example 1: Suppose a population is growing according to the model $P = 800e^{0.05t}$, where t is given in years.
(a) What is the initial size of the population?
(b) How long will it take this population to double?

To estimate the age of dead organic matter, scientists use the carbon dating model ______________________________, which denotes the ratio R of carbon 14 to carbon 12 present at any time t (in years).

Example 2: The ratio of carbon 14 to carbon 12 in a fossil is $R = 10^{-16}$. Find the age of the fossil.

III. Gaussian Models (Page 225)

What you should learn
How to use Gaussian functions to model and solve real-life problems

The Gaussian model is commonly used in probability and statistics to represent populations that are ____________ __________________.

On a bell-shaped curve, the average value for a population is where the ______________________ of the function occurs.

Example 3: Draw the basic form of the graph of a Gaussian model.

y

x

IV. Logistic Growth Models (Page 226)

What you should learn
How to use logistic growth functions to model and solve real-life problems

Give an example of a real-life situation that is modeled by a logistic growth model.

Example 4: Draw the basic form of the graph of a logistic growth model.

y

x

V. Logarithmic Models (Page 227)

What you should learn
How to use logarithmic functions to model and solve real-life problems

Example 5: The number of kitchen widgets y (in millions) demanded each year is given by the model $y = 2 + 3\ \ln(x+1)$, where $x = 0$ represents the year 2000 and $x \geq 0$. Find the year in which the number of kitchen widgets demanded will be 8.6 million.

Homework Assignment

Page(s)

Exercises

Name ____ Date __________

Section 3.6 Nonlinear Models

Objective: In this lesson you learned how to fit exponential, logarithmic, power, and logistic models to sets of data.

I. Classifying Scatter Plots (Page 233)

> ***What you should learn***
> How to classify scatter plots

When faced with a set of data to be modeled, what is a good first step in selecting which type of model will best fit the data?

II. Fitting Nonlinear Models to Data (Pages 234–235)

> ***What you should learn***
> How to use scatter plots and a graphing utility to find models for data and choose the model that best fits a set of data

Describe how to use a graphing utility to fit a nonlinear model to data.

Example 2: Find an appropriate model, either logarithmic or exponential, for the data in the following table.

x	1	3	5	7	9
y	1.120	2.195	4.303	8.433	16.529

III. Modeling With Exponential and Logistic Functions (Pages 236–237)

What you should learn
How to use a graphing utility to find exponential and logistic models for data

Example 3: Find a logistic model for the data in the following table.

x	0	10	15	20	25	30
y	5	27	50	73	88	95

Additional notes

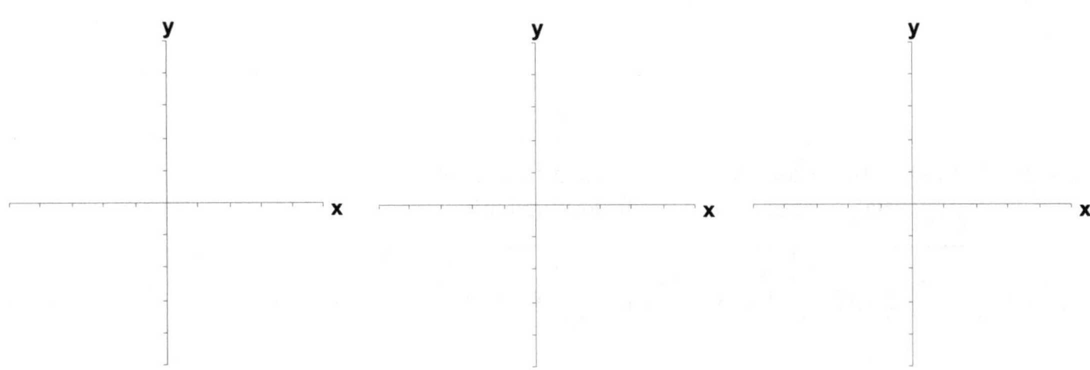

Homework Assignment

Page(s)

Exercises

Name ____ Date __________

Chapter 4 Trigonometric Functions

Section 4.1 Radian and Degree Measure

Objective: In this lesson you learned how to describe an angle and to convert between degree and radian measure.

Important Vocabulary Define each term or concept.

Trigonometry

Central angle of a circle

Complementary angles

Supplementary angles

Degree

I. Angles (Page 254)

What you should learn
How to describe angles

An **angle** is determined by ________________________________
________________________________.

The **initial side** of an angle is ________________________________
________________________________.

The **terminal side** of an angle is ________________________________
________________________________.

The **vertex** of an angle is ________________________________
________________________________.

An angle is in **standard position** when ________________________
____
________________________________.

A **positive angle** is generated by a ________________________
rotation; whereas a **negative angle** is generated by a
________________________ rotation.

If two angles are **coterminal,** then they have ________________
________________________________.

II. Radian Measure (Pages 255–256)

> ***What you should learn***
> How to use radian measure

The measure of an angle is determined by ________________

____.

One **radian** is the measure of a central angle θ that __________

____.

Algebraically this means that $\theta =$ ________________

____.

A central angle of one full revolution (counterclockwise) corresponds to an arc length of $s =$ ____________.

The radian measure of an angle of one full revolution is ___________ radians. A half revolution corresponds to an angle of ___________ radians. Similarly $\frac{1}{4}$ revolution corresponds to an angle of ___________ radians, and $\frac{1}{6}$ revolution corresponds to an angle of ___________ radians.

Angles with measures between 0 and $\pi/2$ radians are ______________ angles. Angles with measures between $\pi/2$ and π radians are ______________ angles.

To find an angle that is coterminal to a given angle θ, ________

____.

Example 1: Find an angle that is coterminal with $\theta = -\pi/8$.

Example 2: Find the supplement of $\theta = \pi/4$.

III. Degree Measure (Pages 257–258)

> ***What you should learn***
> How to use degree measure and convert between degree and radian measure

A full revolution (counterclockwise) around a circle corresponds to ____________ degrees. A half revolution around a circle corresponds to ____________ degrees.

To convert degrees to radians, ______________________________

____.

To convert radians to degrees, ______________________________

____.

Example 3: Convert 120° to radians.

Example 4: Convert $9\pi/8$ radians to degrees.

Example 5: Complete the following table of equivalent degree and radian measures for common angles.

θ (degrees)	0°		45°		90°		270°
θ (radians)		$\pi/6$		$\pi/3$		π	

IV. Linear and Angular Speed (Pages 259–260)

> ***What you should learn***
> How to use angles to model and solve real-life problems

For a circle of radius r, a central angle θ intercepts an arc of length s given by ________________, where θ is measured in radians. Note that if $r = 1$, then $s = \theta$, and the radian measure of θ equals ________________________________.

Consider a particle moving at constant speed along a circular arc of radius r. If s is the length of the arc traveled in time t, then the **linear speed** of the particle is

linear speed = ____________________________________

If θ is the angle (in radian measure) corresponding to the arc length s, then the **angular speed** of the particle is

angular speed = ____________________________________

Example 6: A 6-inch-diameter gear makes 2.5 revolutions per second. Find the angular speed of the gear in radians per second.

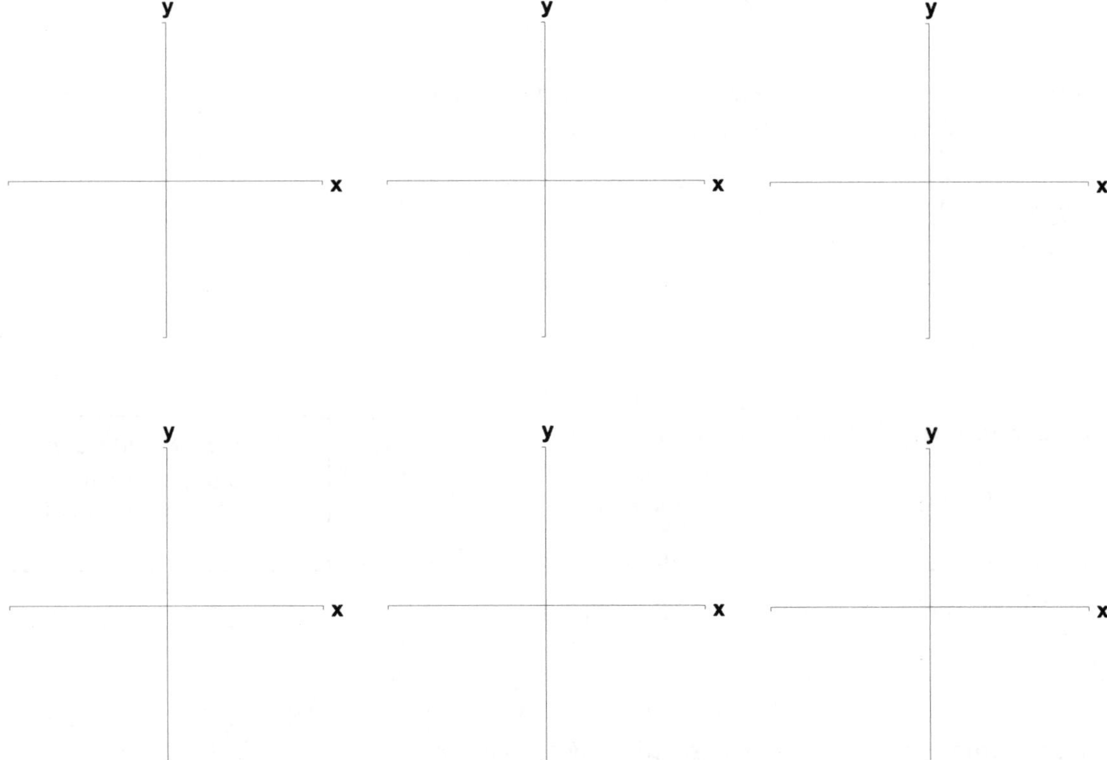

Homework Assignment
Page(s)
Exercises

Name ____ Date ___________

Section 4.2 Trigonometric Functions: The Unit Circle

Objective: In this lesson you learned how to identify a unit circle and describe its relationship to real numbers.

Important Vocabulary	Define each term or concept.
Unit circle	
Periodic	
Period	

I. The Unit Circle (Page 265)

What you should learn
How to identify a unit circle and describe its relationship to real numbers

As the real number line is wrapped around the unit circle, each real number t corresponds to ______________________________ ______________________________.

The real number 2π corresponds to the point ______________ on the unit circle.

Each real number t also corresponds to a __________________ (in standard position) whose radian measure is t. With this interpretation of t, the arc length formula $s = r\theta$ (with $r = 1$) indicates that ____ ____.

II. The Trigonometric Functions (Pages 266–267)

What you should learn
How to evaluate trigonometric functions using the unit circle

The coordinates x and y are two functions of the real variable t. These coordinates can be used to define six trigonometric functions of t. List the abbreviation for each trigonometric function.

Sine	____________	**Cosecant**	____________
Cosine	____________	**Secant**	____________
Tangent	____________	**Cotangent**	____________

Let t be a real number and let (x, y) be the point on the unit circle corresponding to t. Complete the following definitions of the trigonometric functions:

$\sin t =$ ____________ $\cos t =$ ____________

$\tan t =$ ____________ $\cot t =$ ____________

$\sec t =$ ____________ $\csc t =$ ____________

The cosecant function is the reciprocal of the ______________ function. The cotangent function is the reciprocal of the ______________ function. The secant function is the reciprocal of the ______________ function.

Complete the following table showing the correspondence between the real number t and the point (x, y) on the unit circle when the unit circle is divided into eight equal arcs.

t	0	$\pi/4$	$\pi/2$	$3\pi/4$	π	$5\pi/4$	$3\pi/2$	$7\pi/4$
x								
y								

Complete the following table showing the correspondence between the real number t and the point (x, y) on the unit circle when the unit circle is divided into 12 equal arcs.

t	0	$\pi/6$	$\pi/3$	$\pi/2$	$2\pi/3$	$5\pi/6$	π	$7\pi/6$	$4\pi/3$	$3\pi/2$	$5\pi/3$	$11\pi/6$
x												
y												

Example 1: Find the following:

(a) $\cos\dfrac{\pi}{3}$ (b) $\tan\dfrac{3\pi}{4}$ (c) $\csc\dfrac{7\pi}{6}$

III. Domain and Period of Sine and Cosine (Pages 268–269)

> ***What you should learn***
> How to use domain and period to evaluate sine and cosine functions and how to use a calculator to evaluate trigonometric functions

The sine function's domain is ________________________,
and its range is ________________________.
The cosine function's domain is ________________________,
and its range is ________________________.

The period of the sine function is ________________. The period of the cosine function is ________________.

Which trigonometric functions are even functions?

____

Which trigonometric functions are odd functions?

____

Example 2: Evaluate $\sin \frac{31\pi}{6}$

To evaluate the secant function with a calculator, ____________
____.

Example 3: Use a calculator to evaluate (a) $\tan 4\pi/3$, and (b) $\cos 3$.

Additional notes

Homework Assignment

Page(s)

Exercises

Name ____ Date ___________

Section 4.3 Right Triangle Trigonometry

Objective: In this lesson you learned how to evaluate trigonometric functions of acute angles and how to use the fundamental trigonometric identities.

I. The Six Trigonometric Functions (Pages 273–275)

> ***What you should learn***
> How to evaluate trigonometric functions of acute angles and use a calculator to evaluate trigonometric functions

In the right triangle shown below, label the three sides of the triangle relative to the angle labeled θ as (a) the **hypotenuse,** (b) the **opposite side,** and (c) the **adjacent side.**

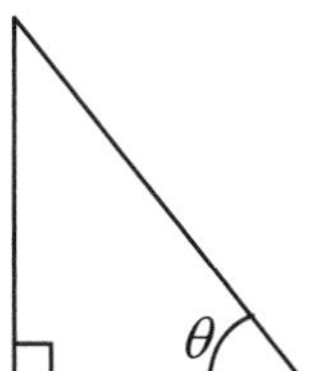

Let θ be an acute angle of a right triangle. Define the six trigonometric functions of the angle θ using opp = the length of the side opposite θ, adj = the length of the side adjacent to θ, and hyp = the length of the hypotenuse.

$\sin\theta =$ ____________ $\cos\theta =$ ____________

$\tan\theta =$ ____________ $\csc\theta =$ ____________

$\sec\theta =$ ____________ $\cot\theta =$ ____________

The cosecant function is the reciprocal of the ______________ function. The cotangent function is the reciprocal of the ______________ function. The secant function is the reciprocal of the ______________ function.

Example 1: In the right triangle below, find $\sin\theta$, $\cos\theta$, and $\tan\theta$.

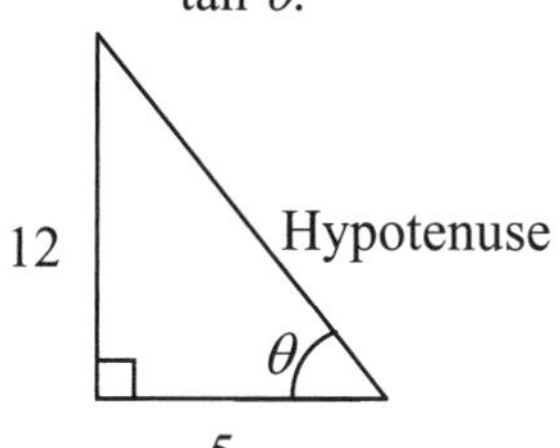

Give the sines, cosines, and tangents of the following special angles:

$\sin 30° = \sin \frac{\pi}{6} =$ ______________________________

$\cos 30° = \cos \frac{\pi}{6} =$ ______________________________

$\tan 30° = \tan \frac{\pi}{6} =$ ______________________________

$\sin 45° = \sin \frac{\pi}{4} =$ ______________________________

$\cos 45° = \cos \frac{\pi}{4} =$ ______________________________

$\tan 45° = \tan \frac{\pi}{4} =$ ______________________________

$\sin 60° = \sin \frac{\pi}{3} =$ ______________________________

$\cos 60° = \cos \frac{\pi}{3} =$ ______________________________

$\tan 60° = \tan \frac{\pi}{3} =$ ______________________________

Cofunctions of complementary angles are ______________. If θ is an acute angle, then:

$\sin(90° - \theta) =$ __________ $\cos(90° - \theta) =$ __________

$\tan(90° - \theta) =$ __________ $\cot(90° - \theta) =$ __________

$\sec(90° - \theta) =$ __________ $\csc(90° - \theta) =$ __________

To use a calculator to evaluate trigonometric functions of angles measured in degrees, ________________________________

____.

Example 2: Use a calculator to evaluate (a) tan 35.4°, and (b) cos 3.25°

II. Trigonometric Identities (Pages 276–277)

> ***What you should learn***
> How to use the fundamental trigonometric identities

List six reciprocal identities:

1)

2)

3)

4)

5)

6)

List two quotient identities:

1)

2)

List three Pythagorean identities:

1)

2)

3)

III. Applications Involving Right Triangles (Pages 278–279)

> ***What you should learn***
> How to use trigonometric functions to model and solve real-life problems

What does it mean to "solve a right triangle?"

An **angle of elevation** is ______________________________

______________________________________.

An **angle of depression** is ____________________________

____________________________________.

Describe a real-life situation in which solving a right triangle would be appropriate or useful.

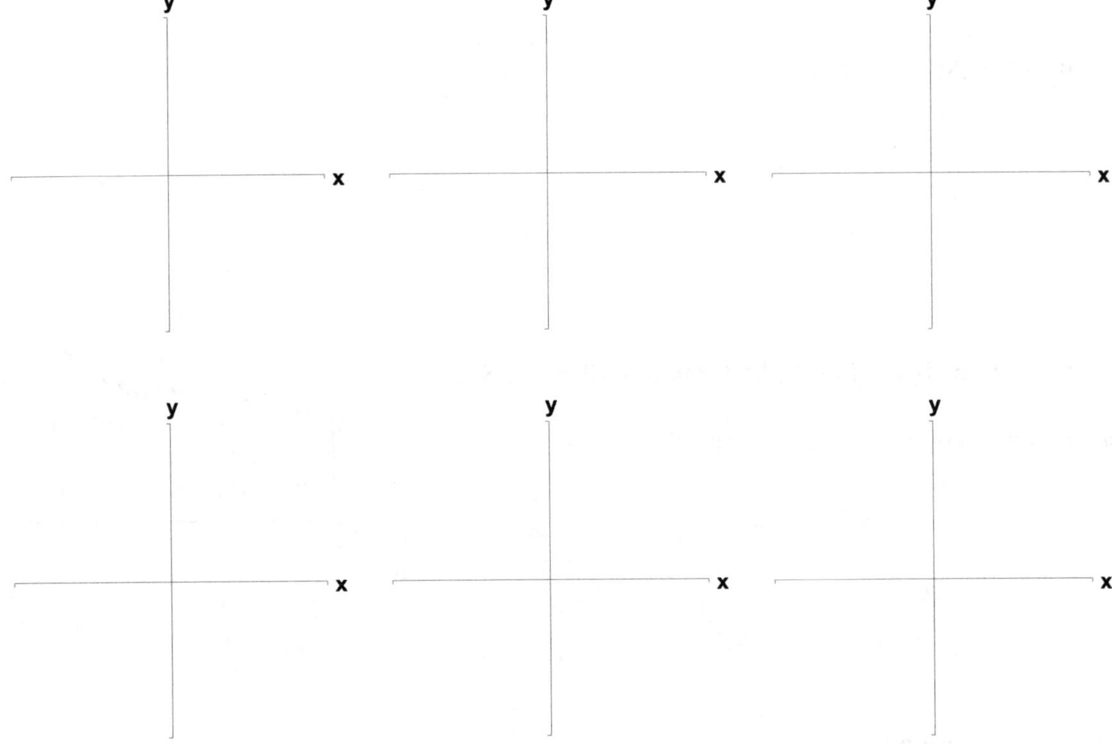

Homework Assignment

Page(s)

Exercises

Name ____ Date ___________

Section 4.4 Trigonometric Functions of Any Angle

Objective: In this lesson you learned how to evaluate trigonometric functions of any angle.

Important Vocabulary	Define each term or concept.
Reference angles	

I. Introduction (Pages 284–285)

> ***What you should learn***
> How to evaluate trigonometric functions of any angle

Let θ be an angle in standard position with (x, y) a point on the terminal side of θ and $r = \sqrt{x^2 + y^2} \neq 0$. Complete the following definitions of the trigonometric functions of any angle:

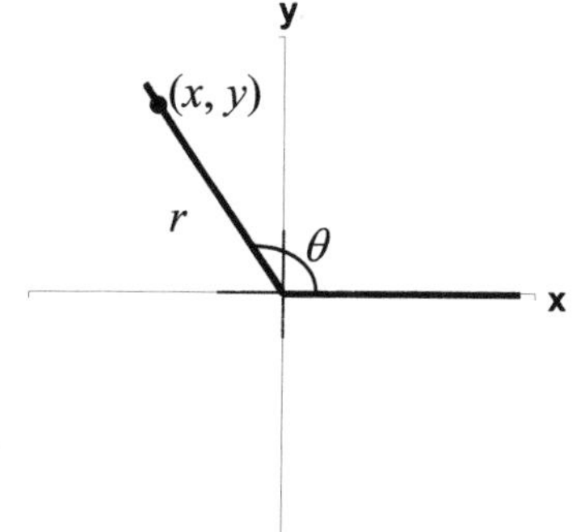

$\sin\theta =$ ____________ $\cos\theta =$ ____________

$\tan\theta =$ ____________ $\cot\theta =$ ____________

$\sec\theta =$ ____________ $\csc\theta =$ ____________

Name the quadrants in which the sine function is positive.

____

Name the quadrants in which the sine function is negative.

____

Name the quadrants in which the cosine function is positive.

____

Name the quadrants in which the cosine function is negative.

____

Name the quadrants in which the tangent function is positive.

____

Name the quadrants in which the tangent function is negative.

____

Example 1: If $\sin\theta = \frac{1}{2}$ and $\tan\theta < 0$, find $\cos\theta$.

II. Reference Angles (Page 286)

> ***What you should learn***
> How to find reference angles

Example 2: Find the reference angle θ' for
(a) $\theta = 210°$ (b) $\theta = 4.1$

III. Trigonometric Functions of Real Numbers (Pages 287–288)

> ***What you should learn***
> How to evaluate trigonometric functions of real numbers

Describe how to find the value of a trigonometric function of any angle θ.

Example 3: Evaluate $\sin \dfrac{11\pi}{6}$.

Example 4: Evaluate $\cos 240°$.

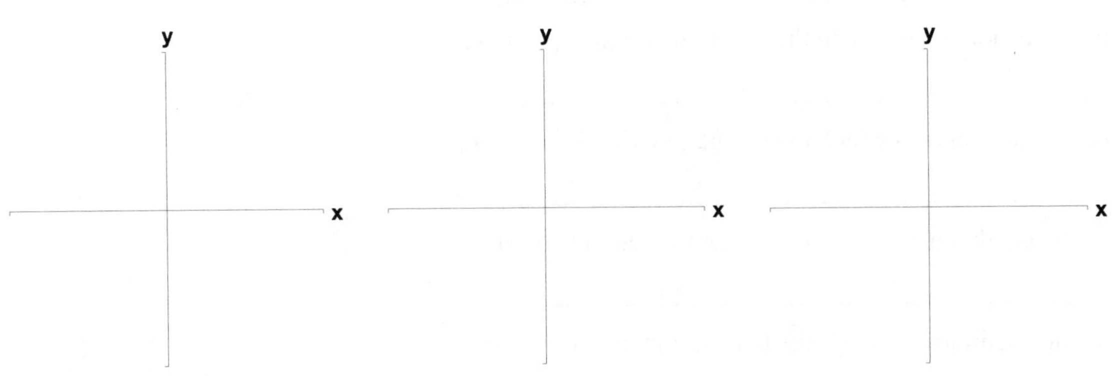

Homework Assignment

Page(s)

Exercises

Name ____ Date __________

Section 4.5 Graphs of Sine and Cosine Functions

Objective: In this lesson you learned how to sketch the graphs of sine and cosine functions and translations of these functions.

Important Vocabulary Define each term or concept.

Sine curve

One cycle

Amplitude

Phase shift

I. Basic Sine and Cosine Curves (Pages 292–293)

What you should learn
How to sketch the graphs of basic sine and cosine functions

For $0 \le x \le 2\pi$, the sine function has its maximum point at ______________, its minimum point at ______________, and its intercepts at ______________________________.

For $0 \le x \le 2\pi$, the cosine function has its maximum points at ______________________, its minimum point at __________, and its intercepts at ______________________________.

Example 1: Sketch the basic sine curve on the interval $[0, 2\pi]$.

y

x

Example 2: Sketch the basic cosine curve on the interval $[0, 2\pi]$.

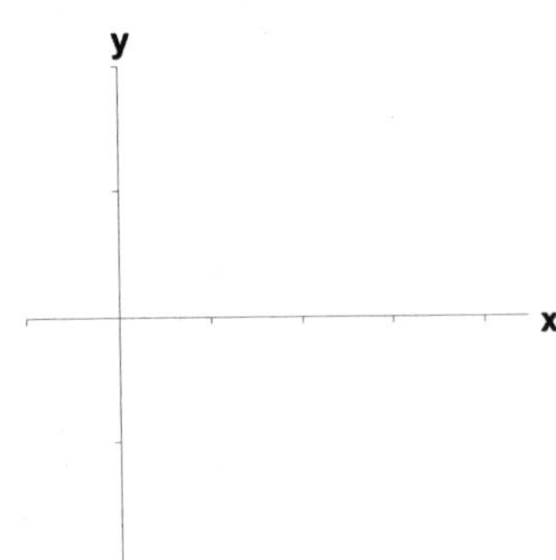

II. Amplitude and Period of Sine and Cosine Curves (Pages 294–295)

> ***What you should learn***
> How to use amplitude and period to help sketch the graphs of sine and cosine functions

The constant factor a in $y = a \sin x$ acts as _______________

____

_______________________________________.

If $|a| > 1$, the basic sine curve is ____________________. If $|a| < 1$, the basic sine curve is _________________. The result is that the graph of $y = a \sin x$ ranges between _________________ instead of between – 1 and 1. The absolute value of a is the __________________ of the function $y = a \sin x$.

The graph of $y = -0.5 \sin x$ is a(n) ____________________ in the x-axis of the graph of $y = 0.5 \sin x$.

Let b be a positive real number. The **period** of $y = a \sin bx$ and $y = a \cos bx$ is __________________. If $0 < b < 1$, the period of $y = a \sin bx$ is __________________ than 2π and represents a ______________________________ of the graph of $y = a \sin bx$.

If $b > 1$, the period of $y = a \sin bx$ is ________________ than 2π and represents a ______________________________ of the graph of $y = a \sin bx$.

Example 3: Find the amplitude and the period of $y = -4\cos 3x$.

Example 4: Find the five key points (intercepts, maximum points, and minimum points) of the graph of $y = -4\cos 3x$.

III. Translations of Sine and Cosine Curves (Pages 296–297)

What you should learn
How to sketch translations of graphs of sine and cosine functions

The constant c in the general equations $y = a\sin(bx - c)$ and $y = a\cos(bx - c)$ creates ______________________________

______________________________.

Comparing $y = a\sin bx$ with $y = a\sin(bx - c)$, the graph of $y = a\sin(bx - c)$ completes one cycle from ____________ to ______________. By solving for x, you can find the interval for one cycle is found to be ___________ to _____________. This implies that the period of $y = a\sin(bx - c)$ is _____________, and the graph of $y = a\sin(bx - c)$ is the graph of $y = a\sin bx$ shifted by the amount _____________.

Example 5: Find the amplitude, period, and phase shift of $y = 2\sin(x - \pi/4)$.

Example 6: Find the five key points (intercepts, maximum points, and minimum points) of the graph of $y = 2\sin(x - \pi/4)$.

The constant d in the equation $y = d + a\sin(bx - c)$ causes a(n) ________________________. For $d > 0$, the shift is ________ ___________. For $d < 0$, the shift is __________________. The graph oscillates about ____________________________.

IV. Mathematical Modeling (Page 298)

> ***What you should learn***
> How to use sine and cosine functions to model real-life data

Describe a real-life situation which can be modeled by a sine or cosine function.

Example 7: Find a trigonometric function to model the data in the following table.

x	0	$\pi/2$	π	$3\pi/2$	2π
y	2	4	2	0	2

Additional notes

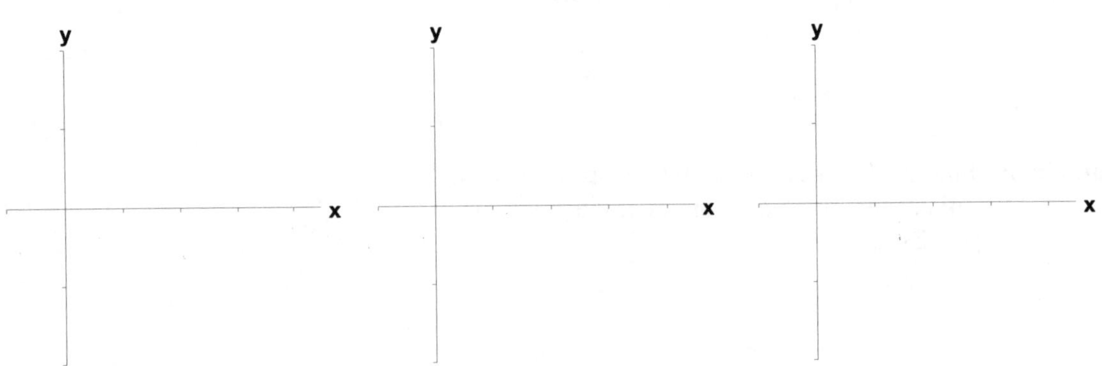

> **Homework Assignment**
>
> Page(s)
>
> Exercises

Name ____ Date ____________

Section 4.6 Graphs of Other Trigonometric Functions

Objective: In this lesson you learned how to sketch the graphs of other trigonometric functions.

Important Vocabulary Define each term or concept.

Damping factor

I. Graph of the Tangent Function (Pages 304–305)

What you should learn
How to sketch the graphs of tangent functions

Because the tangent function is odd, the graph of $y = \tan x$ is symmetric with respect to the ______________. The period of the tangent function is ____________. The tangent function has vertical asymptotes at $x =$ ________________ , where n is an integer. The domain of the tangent function is ______________________________, and the range of the tangent function is ______________________.

Describe how to sketch the graph of a function of the form $y = a\tan(bx - c)$.

II. Graph of the Cotangent Function (Page 306)

What you should learn
How to sketch the graphs of cotangent functions

The period of the cotangent function is ____________. The domain of the cotangent function is ____________________, and the range of the cotangent function is _________________.

The vertical asymptotes of the cotangent function occur at $x =$ ____________, where n is an integer.

III. Graphs of the Reciprocal Functions (Pages 307–308)

What you should learn
How to sketch the graphs of secant and cosecant functions

At a given value of x, the y-coordinate of csc x is the reciprocal of the y-coordinate of ________________.
The graph of $y = \csc x$ is symmetric with respect to the ________________. The period of the cosecant function is ____________. The cosecant function has vertical asymptotes at $x =$ ________________, where n is an integer. The domain of the cosecant function is ____________________, and the range of the cosecant function is ___________________.

At a given value of x, the y-coordinate of sec x is the reciprocal of the y-coordinate of ________________. The graph of $y = \sec x$ is symmetric with respect to the _______________.
The period of the secant function is ________. The secant function has vertical asymptotes, at $x =$ _________________.
The domain of the secant function is _______________________, and the range of the secant function is _____________________.

Describe how to sketch the graph of a secant or cosecant function.

In comparing the graphs of the cosecant and secant functions with those of the sine and cosine functions, note that the "hills" and "valleys" are _____________________.

IV. Damped Trigonometric Graphs (Pages 309–310)

> ***What you should learn***
> How to sketch the graphs of damped trigonometric functions

Explain how to sketch the graph of the damped trigonometric function $y = f(x)\cos x$, where $f(x)$ is the damping factor.

Additional notes

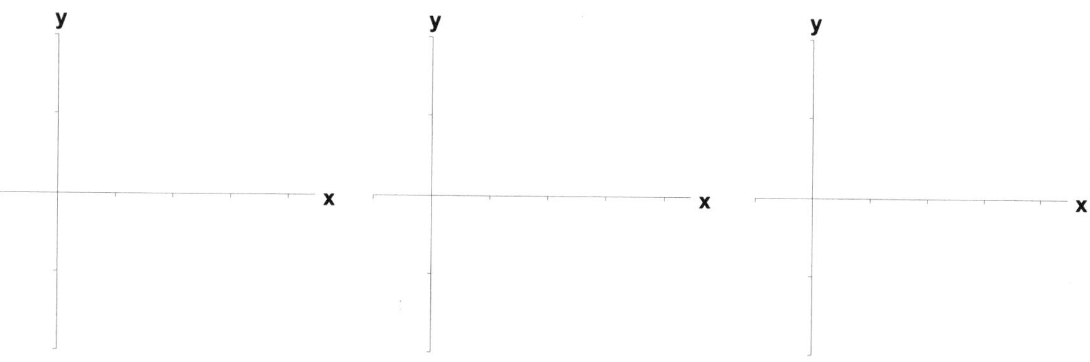

Additional notes

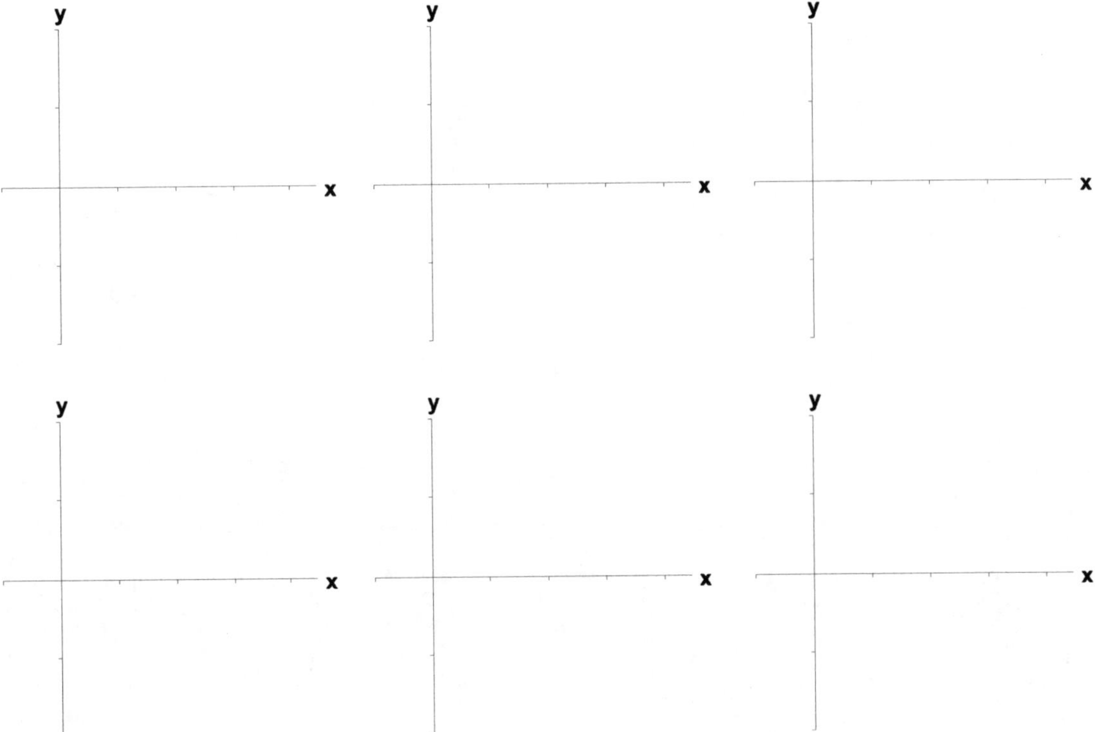

Homework Assignment

Page(s)

Exercises

Name ____ Date ____________

Section 4.7 Inverse Trigonometric Functions

Objective: In this lesson you learned how to evaluate the inverse trigonometric functions and how to evaluate the composition of trigonometric functions.

I. Inverse Sine Function (Pages 315–316)

> ***What you should learn***
> How to evaluate and graph inverse sine functions

Give the definition of the **inverse sine function.**

The domain of $y = \arcsin x$ is ________________. The range of $y = \arcsin x$ is ____________________.

Example 1: Find the exact value: $\arcsin(-1)$.

II. Other Inverse Trigonometric Functions (Pages 317–319)

> ***What you should learn***
> How to evaluate and graph other inverse trigonometric functions

Give the definition of the **inverse cosine function.**

The domain of $y = \arccos x$ is ________________. The range of $y = \arccos x$ is ____________________.

Example 2: Find the exact value: $\arccos \frac{1}{2}$.

Give the definition of the **inverse tangent function.**

The domain of $y = \arctan x$ is ________________. The range of $y = \arctan x$ is ____________________.

Example 3: Find the exact value: $\arctan(\sqrt{3})$.

Example 4: Use a calculator to approximate the value (if possible). Round to four decimal places.
(a) arcos 0.85 (b) arcsin 3.1415

III. Compositions of Functions (Pages 320–321)

What you should learn
How to evaluate compositions of trigonometric functions

State the Inverse Property for the Sine function.

State the Inverse Property for the Cosine function.

State the Inverse Property for the Tangent function.

Example 5: If possible, find the exact value:
(a) arcsin(sin $3\pi/4$) (b) cos(arccos 0)

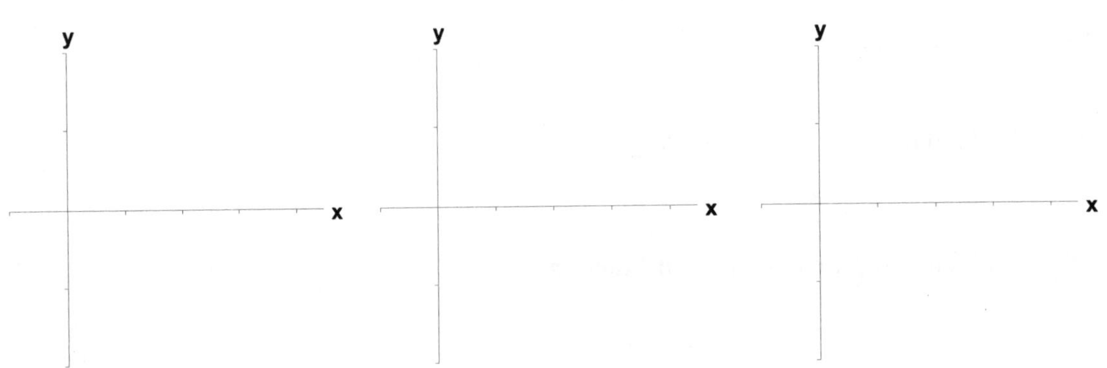

Homework Assignment

Page(s)

Exercises

Name ____ Date __________

Section 4.8 Applications and Models

Objective: In this lesson you learned how to use trigonometric functions to model and solve real-life problems.

I. Applications Involving Right Triangles (Pages 326–327)

> ***What you should learn***
> How to solve real-life problems involving right triangles

Example 1: A ladder leaning against a house reaches 24 feet up the side of the house. The ladder makes a 60° angle with the ground. How far is the base of the ladder from the house? Round your answer to two decimal places.

II. Trigonometry and Bearings (Page 328)

> ***What you should learn***
> How to solve real-life problems involving directional bearings

Used to give directions in surveying and navigation, a **bearing** measures ____

____.

The bearing N 70° E means ______________________________.

Example 2: Write the bearing for the path shown in the diagram at the right.

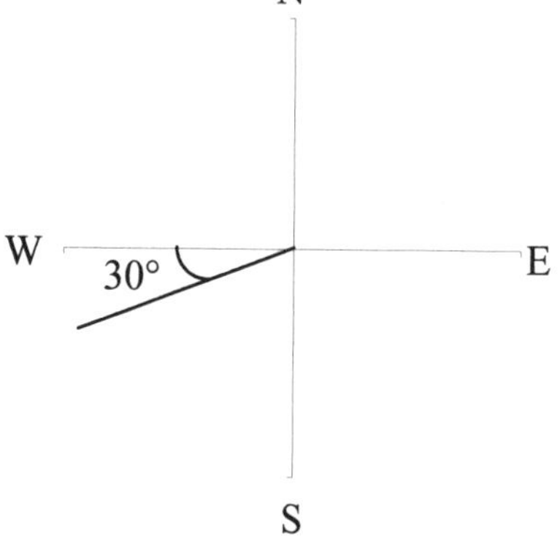

III. Harmonic Motion (Pages 329–331)

> ***What you should learn***
> How to solve real-life problems involving harmonic motion

The vibration, oscillation, or rotation of an object under ideal conditions such that the object's uniform and regular motion can be described by a sine or cosine function is called ____________

____________________.

A point that moves on a coordinate line is said to be in **simple harmonic motion** if ______________________________

____

____.

The simple harmonic motion has amplitude ___________,
period
___________, and frequency _________________.

Example 3: Given the equation for simple harmonic motion $d = 3\sin\frac{t}{2}$, find:

(a) the maximum displacement,
(b) the frequency of the simple harmonic motion, and
(c) the period of the simple harmonic motion.

y x

y x

y x

Homework Assignment

Page(s)

Exercises

Name ______________________________ Date __________

Chapter 5 Analytic Trigonometry

Section 5.1 Using Fundamental Identities

Objective: In this lesson you learned how to use fundamental trigonometric identities to evaluate trigonometric functions and simplify trigonometric expressions.

I. Introduction (Page 350)

What you should learn How to recognize and write the fundamental trigonometric identities

Name four ways in which the fundamental trigonometric identities can be used:

1)

2)

3)

4)

The Fundamental Trigonometric Identities

List six reciprocal identities:

1)

2)

3)

4)

5)

6)

List two quotient identities:

1)

2)

List three Pythagorean identities:

1)

2)

3)

List six cofunction identities:

1)

2)

3)

4)

5)

6)

List six even/odd identities:

1)

2)

3)

4)

5)

6)

II. Using the Fundamental Identities (Pages 351–353)

What you should learn
How to use the fundamental trigonometric identities to evaluate trigonometric functions, simplify trigonometric expressions, and rewrite trigonometric expressions

Example 1: Explain how to use the fundamental trigonometric identities to find the value of tan u given that $\sec u = 2$.

Example 2: Explain how to use the fundamental trigonometric identities to simplify $\sec x - \tan x \sin x$.

Example 3: Explain how to use a graphing utility to verify whether $\sec x \sin^3 x + \sin x \cos x = \tan x$ is an identity.

Homework Assignment

Page(s)

Exercises

Name ____ Date ___________

Section 5.2 Verifying Trigonometric Identities

Objective: In this lesson you learned how to verify trigonometric identities.

I. Verifying Trigonometric Identities (Pages 357–361)

What you should learn
How to verify trigonometric identities

The key to both verifying identities and solving equations is ____
____
____.

An identity is _______________________________
____.

Complete the following list of guidelines for verifying trigonometric identities:

1)

2)

3)

4)

5)

Example 1: Describe a strategy for verifying the identity $\sin\theta\tan\theta + \cos\theta = \sec\theta$. Then verify the identity.

Example 2: Describe a strategy for verifying the identity $\sin^2 x(\csc x - 1)(\csc x + 1) = 1 - \sin^2 x$. Then verify the identity.

Example 3: Verify the identity $\cot^5 \alpha = \cot^3 \alpha \csc^2 \alpha - \cot^3 \alpha$.

Additional notes

Homework Assignment

Page(s)

Exercises

Name ____ Date __________

Section 5.3 Solving Trigonometric Equations

Objective: In this lesson you learned how to use standard algebraic techniques and inverse trigonometric functions to solve trigonometric equations.

I. Introduction (Pages 365–367)

> ***What you should learn***
> How to use standard algebraic techniques to solve trigonometric equations

To solve a trigonometric equation, ______________________________

____.

The preliminary goal in solving trigonometric equations is _____

____.

How many solutions does the equation $\sec x = 2$ have? Explain.

Example 1: Solve $2\cos^2 x - 1 = 0$.

To solve an equation in which two or more trigonometric functions occur, ______________________________

____

______________________________.

II. Equations of Quadratic Type (Pages 368–369)

> ***What you should learn***
> How to solve trigonometric equations of quadratic type

Give an example of a trigonometric equation of quadratic type.

To solve a trigonometric equation of quadratic type,

____

______________________________.

Example 2: Solve $\tan^2 x + 2\tan x = -1$.

Care must be taken when squaring each side of a trigonometric equation to obtain a quadratic because ____________________

____

____

____.

III. Functions Involving Multiple Angles (Page 370)

> ***What you should learn***
> How to solve trigonometric equations involving multiple angles

Give an example of a trigonometric function of multiple angles.

Example 3: Solve $\sin 4x = \frac{\sqrt{2}}{2}$.

IV. Using Inverse Functions (Page 371–372)

> ***What you should learn***
> How to use inverse trigonometric functions to solve trigonometric equations

Example 4: Use inverse functions to solve the equation $\tan^2 x + 4\tan x + 4 = 0$.

Homework Assignment
Page(s)
Exercises

Name ____ Date ___________

Section 5.4 Sum and Difference Formulas

Objective: In this lesson you learned how to use sum and difference formulas to rewrite and evaluate trigonometric functions.

I. Using Sum and Difference Formulas (Pages 377–380)

> ***What you should learn***
> How to use sum and difference formulas to evaluate trigonometric functions, verify identities, and solve trigonometric equations

List the sum and difference formulas for sine, cosine, and tangent.

Example 1: Use a sum or difference formula to find the exact value of tan 255°.

Example 2: Find the exact value of cos 95° cos 35° + sin 95° sin 35°.

A **reduction formula** is ______________________________

____

____.

Example 3: Derive a reduction formula for $\sin\left(t + \frac{\pi}{2}\right)$.

Example 4: Find all solutions of $\cos(x - \frac{\pi}{3}) + \cos(x + \frac{\pi}{3}) = 1$ in the interval $[0, 2\pi)$.

Additional notes

Homework Assignment

Page(s)

Exercises

Name ____ Date ___________

Section 5.5 Multiple-Angle and Product-to-Sum Formulas

Objective: In this lesson you learned how to use multiple-angle formulas, power-reducing formulas, half-angle formulas, and product-to-sum formulas to rewrite and evaluate trigonometric functions.

I. Multiple-Angle Formulas (Pages 384–385)

> ***What you should learn***
> How to use multiple-angle formulas to rewrite and evaluate trigonometric functions

The most commonly used multiple-angle formulas are the

_________________________, which are listed below:

$\sin 2u =$ ____________________

$\cos 2u =$ ____________________

$=$ ____________________

$=$ ____________________

$\tan 2u =$ ____________________

To obtain other multiple-angle formulas, ________________

____

____

______________________________________.

Example 1: Use multiple-angle formulas to express $\cos 3x$ in terms of $\cos x$.

II. Power-Reducing Formulas (Page 386)

> ***What you should learn***
> How to use power-reducing formulas to rewrite and evaluate trigonometric functions

The double-angle formulas can be used to obtain the

____.

The power-reducing formulas are:

$\sin^2 u =$ ____________________

$\cos^2 u =$ ____________________

$\tan^2 u =$ ______________________________

III. Half-Angle Formulas (Page 387)

> ***What you should learn***
> How to use half-angle formulas to rewrite and evaluate trigonometric functions

List the **half-angle formulas:**

$\sin \frac{u}{2} =$ ____________________

$\cos \frac{u}{2} =$ ____________________

$\tan \frac{u}{2} =$ ____________________ $=$ ____________________

The signs of sin ($u/2$) and cos ($u/2$) depend on ______________

____.

Example 2: Find the exact value of tan 15°.

IV. Product-to-Sum Formulas (Pages 388–389)

> ***What you should learn***
> How to use product-to-sum and sum-to-product formulas to rewrite and evaluate trigonometric functions

The **product-to-sum formulas** are used in calculus to ________

____

____.

The product-to-sum formulas are:

$\sin u \sin v =$ ______________________________

$\cos u \cos v =$ ______________________________

$\sin u \cos v =$ ______________________________

$\cos u \sin v =$ ______________________________

Example 3: Write $\cos 3x \cos 2x$ as a sum or difference.

The **sum-to-product formulas** can be used to ______________

____.

The sum-to-product formulas are:

$\sin u + \sin v =$ _____________________________________

$\sin u - \sin v =$ _____________________________________

$\cos u + \cos v =$ _____________________________________

$\cos u - \cos v =$ _____________________________________

Example 4: Write $\cos 4x + \cos 2x$ as a sum or difference.

Additional notes

Additional notes

Homework Assignment

Page(s)

Exercises

Name ____ Date __________

Chapter 6 Additional Topics in Trigonometry

Section 6.1 Law of Sines

Objective: In this lesson you learned how to use the Law of Sines to solve oblique triangles and how to find the areas of oblique triangles.

Important Vocabulary	Define each term or concept.
Oblique triangle	

I. Introduction (Pages 404–405)

> ***What you should learn***
> How to use the Law of Sines to solve oblique triangles (AAS or ASA)

State the **Law of Sines.**

To solve an oblique triangle, you need to know the measure of at least one side and any two other parts of the triangle. Describe two cases that can be solved using the Law of Sines.

Example 1: For the triangle shown at the right, $A = 31.6°$, $C = 42.9°$, and $a = 10.4$ meters. Find the length of side c.

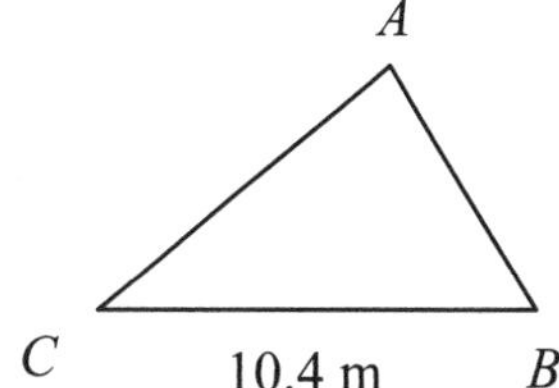

II. The Ambiguous Case (SSA) (Pages 406–407)

> ***What you should learn***
> How to use the Law of Sines to solve oblique triangles (SSA)

If two sides and one opposite angle of an oblique triangle are given, ________________ possible situations can occur, which are:

Example 2: For a triangle having $A = 25°$, $b = 54$ feet, and $a = 26$ feet, how many solutions are possible?

Example 3: For the triangle shown at the right, $A = 110°$, $c = 16$ centimeters, and $a = 25$ centimeters. Find the length of side b.

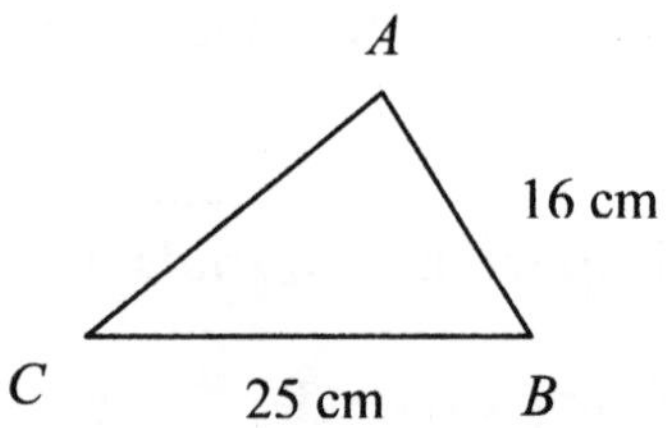

III. Area of an Oblique Triangle (Pages 408–409)

What you should learn
How to find areas of oblique triangles and use the Law of Sines to model and solve real-life problems

The area of any triangle is ______________ the product of the lengths of two sides times the sine of ____________________.
That is,

Area = ____

Example 4: Find the area of a triangle having two sides of lengths 30 feet and 48 feet and an included angle of 40°.

Describe a real-life situation in which the Law of Sines could be used.

Homework Assignment

Page(s)

Exercises

Name ____ Date __________

Section 6.2 Law of Cosines

Objective: In this lesson you learned how to use the Law of Cosines to solve oblique triangles and to use Heron's Formula to find the area of a triangle.

I. Introduction (Pages 413–414)

> ***What you should learn***
> How to use the Law of Cosines to solve oblique triangles (SSS or SAS)

State the **Law of Cosines**.

Example 1: Using the triangle shown at the right, find angle A.

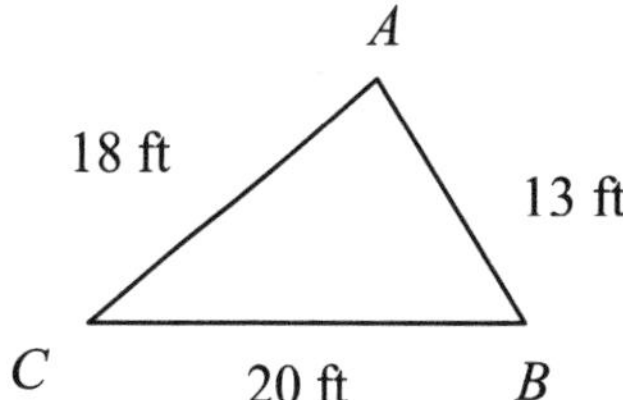

When given the lengths of all three sides of a triangle and asked to find all three angles, which angle should be found first? Why?

Example 2: In the triangle shown at the right, if $A = 62°$, find the length of side a.

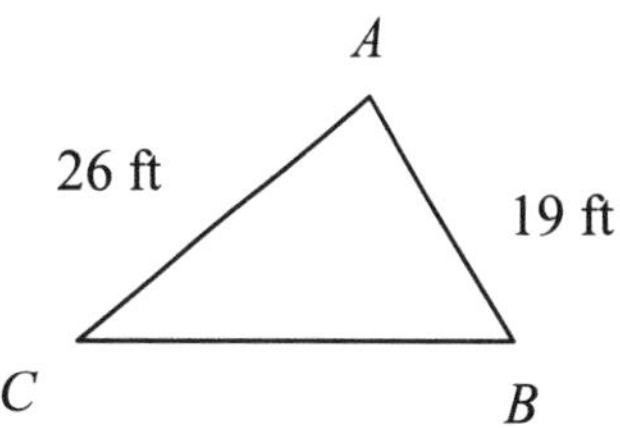

II. Applications (Page 415)

> ***What you should learn***
> How to use the Law of Cosines to model and solve real-life problems

Describe a real-life situation in which the Law of Cosines could be used.

III. Heron's Area Formula (Page 416)

What you should learn
How to use Heron's Area Formula to find areas of triangles

Heron's Area Formula states that given any triangle with sides of length a, b, and c, the area of the triangle is:

Area = $\sqrt{}$ ______________________________

where s = ____________________.

Example 3: Find the area of a triangle having sides of length $a = 14$ cm, $b = 21$ cm, and $c = 27$ cm.

Additional notes

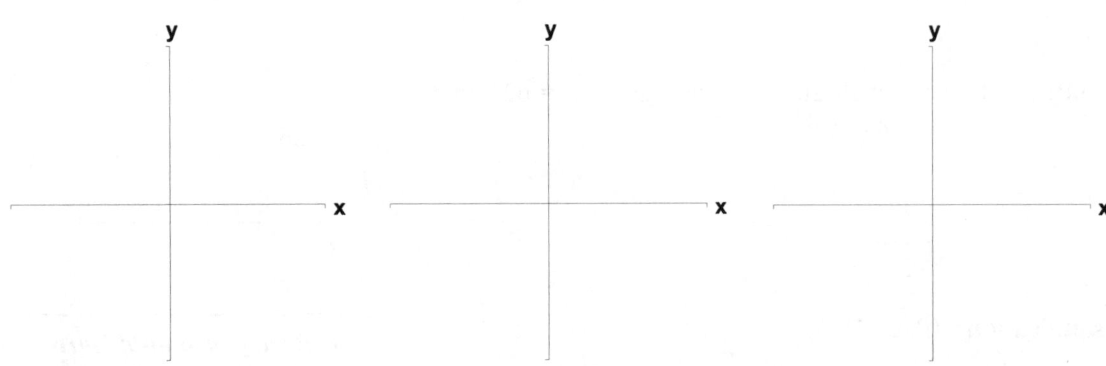

Homework Assignment

Page(s)

Exercises

Name ____ Date ___________

Section 6.3 Vectors in the Plane

Objective: In this lesson you learned how to represent vectors as directed line segments, perform mathematical operations on vectors, and find direction angles of vectors.

Important Vocabulary	Define each term or concept.
Vector v in the plane	
Standard position	
Zero vector	
Unit vector	
Standard unit vectors	
Direction angle	

I. Introduction (Page 420)

> ***What you should learn***
> How to represent vectors as directed line segments

A **directed line segment** $\overrightarrow{PQ}$, has ____________________ P and ______________________ Q.

The **magnitude,** or _______________, of the directed line segment $\overrightarrow{PQ}$, is denoted by _______________ and can be found by ____.

II. Component Form of a Vector (Page 421)

> ***What you should learn***
> How to write the component forms of vectors

A vector whose initial point is at the origin (0, 0) can be uniquely represented by the coordinates of its terminal point (v_1, v_2). This is the ____, written $\mathbf{v} = \langle v_1, v_2 \rangle$, where v_1 and v_2 are the _________________ of **v**.

The component form of the vector with initial point $P = (p_1, p_2)$ and terminal point $Q = (q_1, q_2)$ is

$\overrightarrow{PQ} =$ ______________________ $=$ _____________ $= \mathbf{v}$.

The **magnitude** (or length) of **v** is:

$\|\mathbf{v}\| = \sqrt{\_\_\_\_\_\_\_\_\_\_\_\_\_\_\_\_\_\_\_\_} = \sqrt{\_\_\_\_\_\_\_\_\_\_}$

Example 1: Find the component form and magnitude of the vector **v** that has (1, 7) as its initial point and (4, 3) as its terminal point.

III. Vector Operations (Pages 422–423)

> ***What you should learn***
> How to perform basic vector operations and represent vectors graphically

Geometrically, the product of a vector **v** and a scalar k is ______ ____.

If k is positive, $k\mathbf{v}$ has the ____________ direction as **v**, and if k is negative, $k\mathbf{v}$ has the ____________ direction.

To add two vectors **u** and **v** geometrically, _______________ ____ ____ ____.

This technique is called the _________________________ for vector addition because the vector **u** + **v**, often called the _________________ of vector addition, is _______________ ____.

Let $\mathbf{u} = \langle u_1, u_2 \rangle$ and $\mathbf{v} = \langle v_1, v_2 \rangle$ be vectors and let k be a scalar (a real number). The **sum** of **u** and **v**, is the vector

$\mathbf{u} + \mathbf{v} =$ ______________________

The **scalar multiple** of k times **u**, is the vector:

$k\mathbf{u} =$ __________________________

Example 2: Let $\mathbf{u} = \langle 1, 6 \rangle$ and $\mathbf{v} = \langle -4, 2 \rangle$. Find:
(a) $3\mathbf{u}$
(b) $\mathbf{u} + \mathbf{v}$

Let **u**, **v**, and **w** be vectors and c and d be scalars. Complete the following properties of vector addition and scalar multiplication:

1. $\mathbf{u} + \mathbf{v} =$ ______________
2. $(\mathbf{u} + \mathbf{v}) + \mathbf{w} =$ ___________________
3. $\mathbf{u} + \mathbf{0} =$ ___________
4. $\mathbf{u} + (-\mathbf{u}) =$ ____________
5. $c(d\mathbf{u}) =$ ______________
6. $(c + d)\mathbf{u} =$ _______________
7. $c(\mathbf{u} + \mathbf{v}) =$ _______________
8. $1(\mathbf{u}) =$ ___________ , $0(\mathbf{u}) =$ ___________
9. $\|c\mathbf{v}\| =$ __________________

IV. Unit Vectors (Pages 424–425)

> ***What you should learn***
> How to write vectors as linear combinations of unit vectors

To find a unit vector **u** that has the same direction as a given nonzero vector **v**, ____________________________________

____.

In this case, the vector **u** is called a _______________________

____________________.

Example 3: Find a unit vector in the direction of $\mathbf{v} = \langle -8, 6 \rangle$.

Let $\mathbf{v} = \langle v_1, v_2 \rangle$. Then the standard unit vectors can be used to represent **v** as $\mathbf{v} =$ ________________ , where the scalar v_1 is called the ______________________________ and the scalar v_2 is called the ____________________________. The vector sum $v_1\mathbf{i} + v_2\mathbf{j}$ is called a ________________________ of the vectors **i** and **j**.

Example 4: Let $\mathbf{v} = \langle -5, 3 \rangle$. Write **v** as a linear combination of the standard unit vectors **i** and **j**.

Example 5: Let $\mathbf{v} = 3\mathbf{i} - 4\mathbf{j}$ and $\mathbf{w} = 2\mathbf{i} + 9\mathbf{j}$. Find $\mathbf{v} + \mathbf{w}$.

V. Direction Angles (Page 426)

> ***What you should learn***
> How to find the direction angles of vectors

If **u** is a unit vector and θ is the angle (measured counterclockwise) from the positive x-axis to **u**, the terminal point of **u** lies on the unit circle and

$\mathbf{u} = \langle x, y \rangle =$ ____________________ = ________________________

Now, if $\mathbf{v} = a\mathbf{i} + b\mathbf{j}$ is any vector that makes an angle θ with the positive x-axis, then it has the same direction as **u** and

$\mathbf{v} =$ ________________________ = ________________________

Because $\mathbf{v} = a\mathbf{i} + b\mathbf{j}$, then the direction angle θ for **v** can be determined from $\tan\theta =$ ____________.

Example 6: Let $\mathbf{v} = -4\mathbf{i} + 5\mathbf{j}$. Find the direction angle for **v**.

VI. Applications of Vectors (Pages 427–428)

> ***What you should learn***
> How to use vectors to model and solve real-life problems

Describe several real-life applications of vectors.

Homework Assignment

Page(s)

Exercises

Name ____ Date __________

Section 6.4 Vectors and Dot Products

Objective: In this lesson you learned how to find the dot product of two vectors and use properties of the dot product.

Important Vocabulary	Define each term or concept.
Angle between two nonzero vectors	
Orthogonal vectors	
Vector components	

I. The Dot Product of Two Vectors (Page 434)

> ***What you should learn***
> How to find the dot product of two vectors and use the properties of the dot product

The **dot product** of $\mathbf{u} = \langle u_1, u_2 \rangle$ and $\mathbf{v} = \langle v_1, v_2 \rangle$ is

______________________. This product yields a __________.

Let **u**, **v**, and **w** be vectors in the plane or in space and let c be a scalar. Complete the following properties of the dot product:

1. $\mathbf{u} \bullet \mathbf{v} =$ ______________
2. $\mathbf{0} \bullet \mathbf{v} =$ ____________
3. $\mathbf{u} \bullet (\mathbf{v} + \mathbf{w}) =$ ____________________
4. $\mathbf{v} \bullet \mathbf{v} =$ _______________
5. $c(\mathbf{u} \bullet \mathbf{v}) =$ ______________ = _______________

Example 1: Find the dot product: $\langle 5, -4 \rangle \bullet \langle 9, -2 \rangle$.

II. The Angle Between Two Vectors (Pages 435–436)

> ***What you should learn***
> How to find the angle between two vectors and determine whether two vectors are orthogonal

If θ is the angle between two nonzero vectors **u** and **v**, then

____________________________________.

Example 2: Find the angle between $\mathbf{v} = \langle 5, -4 \rangle$ and $\mathbf{w} = \langle 9, -2 \rangle$.

An alternative way to calculate the dot product between two vectors **u** and **v**, given the angle θ between them, is

______________________________.

Two vectors **u** and **v** are orthogonal if ________________.

Example 3: Are the vectors $\mathbf{u} = \langle 1, -4 \rangle$ and $\mathbf{v} = \langle 6, 2 \rangle$ orthogonal?

III. Finding Vector Components (Pages 437–438)

> ***What you should learn***
> How to write vectors as the sums of two vector components

Let **u** and **v** be nonzero vectors such that $\mathbf{u} = \mathbf{w}_1 + \mathbf{w}_2$, where $\mathbf{w}_1$ and $\mathbf{w}_2$ are orthogonal and $\mathbf{w}_1$ is parallel to (or a scalar multiple of) **v**. The vectors $\mathbf{w}_1$ and $\mathbf{w}_2$ are called ____________________ ____________. The vector $\mathbf{w}_1$ is the **projection** of **u** onto **v** and is denoted by ____________________. The vector $\mathbf{w}_2$ is given by ____________________ .

Let **u** and **v** be nonzero vectors. The projection of **u** onto **v** is given by $\text{proj}_{\mathbf{v}}\,\mathbf{u} =$ ______________________________.

IV. Work (Page 439)

> ***What you should learn***
> How to use vectors to find the work done by a force

The work W done by a constant force **F** as its point of application moves along the vector $\overrightarrow{PQ}$ is given by either of the following:

1.

2.

Homework Assignment
Page(s)
Exercises

Name ____ Date ___________

Section 6.5 Trigonometric Form of a Complex Number

Objective: In this lesson you learned how to multiply and divide complex numbers written in trigonometric form and how to find powers and *n*th roots of complex numbers.

Important Vocabulary	Define each term or concept.
***n*th roots of unity**	

I. The Complex Plane (Page 443)

> ***What you should learn***
> How to plot complex numbers in the complex plane and find absolute values of complex numbers

The **absolute value of the complex number** $a + bi$ is defined as

____.

The absolute value of the complex number $z = a + bi$ is

given by $|a + bi| = \sqrt{\_\_\_\_\_\_\_\_\_\_}$.

II. Trigonometric Form of a Complex Number (Pages 444–445)

> ***What you should learn***
> How to write trigonometric forms of complex numbers

The **trigonometric form of the complex number** $z = a + bi$ is

$z =$ ________________________,

where $a =$ ________________,

$b =$ ________________,

$r = \sqrt{\_\_\_\_\_\_\_\_\_\_}$, and

$\tan \theta =$ ____________.

The number r is the ______________ of z, and θ is called an

________________ of z.

The trigonometric form of a complex number is also called the

__________________.

III. Multiplication and Division of Complex Numbers (Pages 446–447)

What you should learn
How to multiply and divide complex numbers written in trigonometric form

Let $z_1 = r_1(\cos \theta_1 + i \sin \theta_1)$ and $z_2 = r_2(\cos \theta_2 + i \sin \theta_2)$ be complex numbers. Then:

$z_1 z_2 =$ ______________________________

$z_1/z_2 =$ ______________________________

Describe how to multiply two complex numbers.

Describe how to divide two complex numbers.

IV. Powers of Complex Numbers (Page 448)

What you should learn
How to use DeMoivre's Theorem to find powers of complex numbers

State **DeMoivre's Theorem**.

IV. Roots of Complex Numbers (Pages 449–451)

What you should learn
How to find *n*th roots of complex numbers

The complex number $u = a + bi$ is an ***n*th root** of the complex number z if ______________________.

For a positive integer n, the complex number $z = r(\cos \theta + i \sin \theta)$ has ______________________________ given by $\sqrt[n]{r}\left(\cos \frac{\theta + 2\pi k}{n} + i \sin \frac{\theta + 2\pi k}{n}\right)$, where $k = 0, 1, 2, \ldots, n - 1$.

Homework Assignment

Page(s)

Exercises

Name ______________________________ Date __________

Chapter 7 Linear Systems and Matrices

Section 7.1 Solving Systems of Equations

Objective: In this lesson you learned how to solve systems of equations by substitution and by graphing and how to use systems of equations to model and solve real-life problems.

Important Vocabulary Define each term or concept.

Systems of equations

Solution of a system of equations (in two variables)

Method of substitution

Point of intersection

Break-even point

I. The Methods of Substitution and Graphing (Pages 470–474)

What you should learn
How to use the methods of substitution and graphing to solve systems of equations in two variables

To check that the ordered pair (– 3, 4) is the solution of a system of two equations, ______________________________

______________________________.

List the steps necessary for solving a system of two equations in x and y using the method of substitution.

The solution of a system of equations corresponds to the

____ of the

graphs of the equations in the system.

List the necessary steps for using the method of graphing to solve a system of two equations in x and y.

Explain what is meant by back-substitution.

Example 1: Solve the system of equations using the method of substitution.

$$\begin{cases} 2x + y = 2 \\ x - 2y = -9 \end{cases}$$

Example 2: Solve the system of equations using the method of graphing.

$$\begin{cases} x^2 - y = 5 \\ -x + y = -3 \end{cases}$$

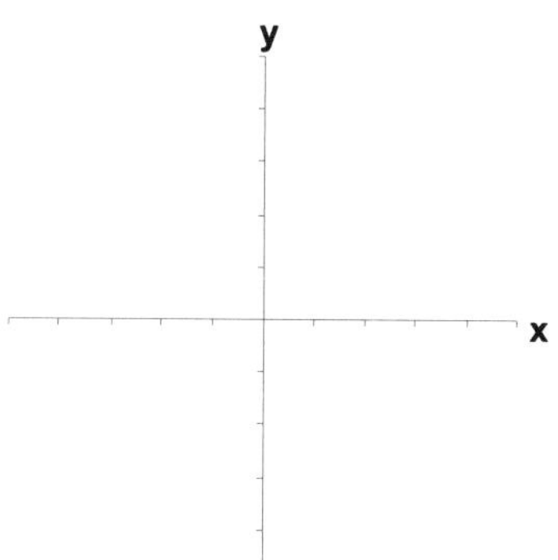

II. Application (Page 475)

What you should learn
How to use systems of equations to model and solve real-life problems

The total cost C of producing x units of a product typically has two components: ____________________________________.

In break-even analysis, the break-even point corresponds to the ______________________________ of the cost and revenue curves.

Break-even analysis can also be approached from the point of view of profit. In this case, consider the profit function, which is ________________. The break-even point occurs when profit equals __________.

Example 3: The cost of producing x units is $C = 1.5x + 15{,}000$ and the revenue obtained by selling x units is $R = 5x$. How many items should be sold to break even?

Additional notes

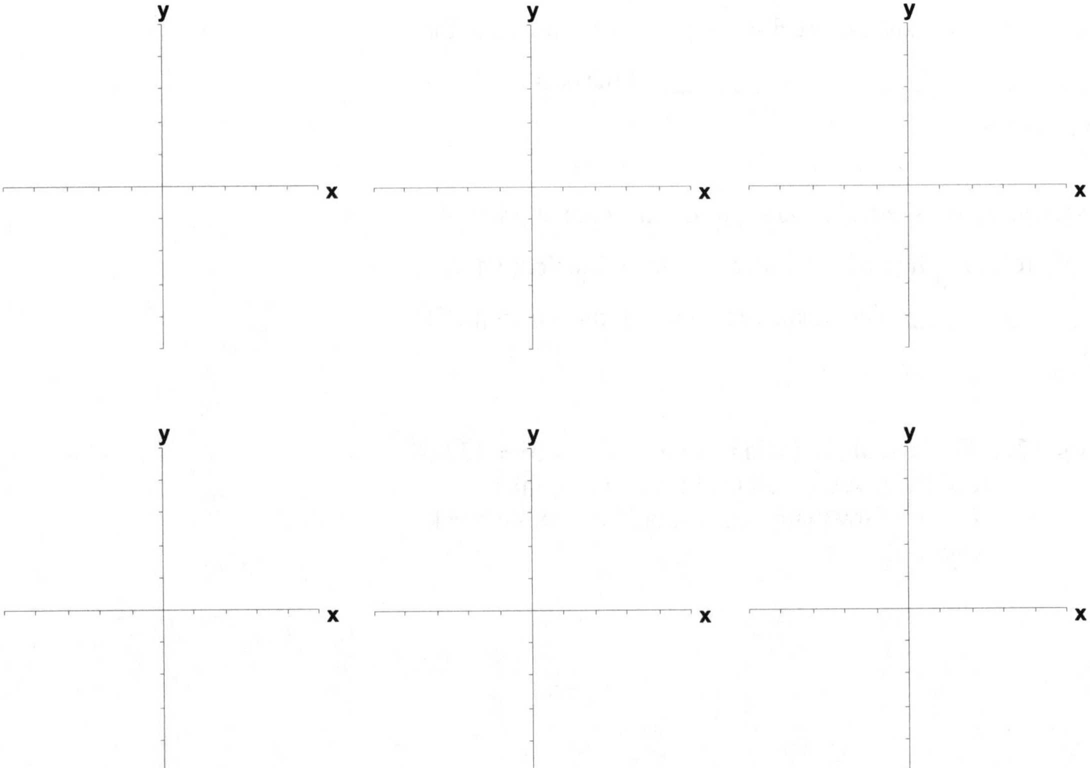

Homework Assignment

Page(s)

Exercises

Name ____ Date ____________

Section 7.2 Systems of Linear Equations in Two Variables

Objective: In this lesson you learned how to solve a system of equations by elimination and how to use systems of equations to model and solve real-life problems.

Important Vocabulary Define each term or concept.

Method of elimination

Equivalent systems

Consistent system

Inconsistent system

I. The Method of Elimination (Pages 480–481)

What you should learn
How to use the method of elimination to solve systems of linear equations in two variables

List the steps necessary for solving a system of two linear equations in x and y using the method of elimination.

The operations that can be performed on a system of linear equations to produce an equivalent system are:

(1)

(2)

(3)

Example 1: Describe a strategy for solving the system of linear equations using the method of elimination.

$$\begin{cases} 3x + y = 9 \\ 4x - 2y = -1 \end{cases}$$

Example 2: Solve the system of linear equations using the method of elimination.

$$\begin{cases} 4x + y = -3 \\ x - 3y = 9 \end{cases}$$

II. Graphical Interpretation of Two-Variable Systems (Pages 482–483)

> ***What you should learn***
> How to graphically interpret the number of solutions of a system of linear equations in two variables

If a system of linear equations has two different solutions, it must have ________________________________ solutions.

For a system of two linear equations in two variables, list the possible number of solutions the system can have and give a graphical interpretation of the solutions.

If a false statement such as 9 = 0 is obtained while solving a system of linear equations using the method of elimination, then the system has ________________________.

If a statement that is true for all values of the variables, such as 0 = 0, is obtained while solving a system of linear equations using the method of elimination, then the system has

____.

Example 3: Is the following system consistent or inconsistent? How many solutions does the system have?

$$\begin{cases} x - 3y = 2 \\ -4x + 12y = 8 \end{cases}$$

III. Application (Page 484)

> ***What you should learn***
> How to use systems of linear equations in two variables to model and solve real-life problems

When may a system of linear equations be an appropriate mathematical model for solving a real-life application?

Give an example of a real-life application that could be solved with a system of linear equations.

Additional notes

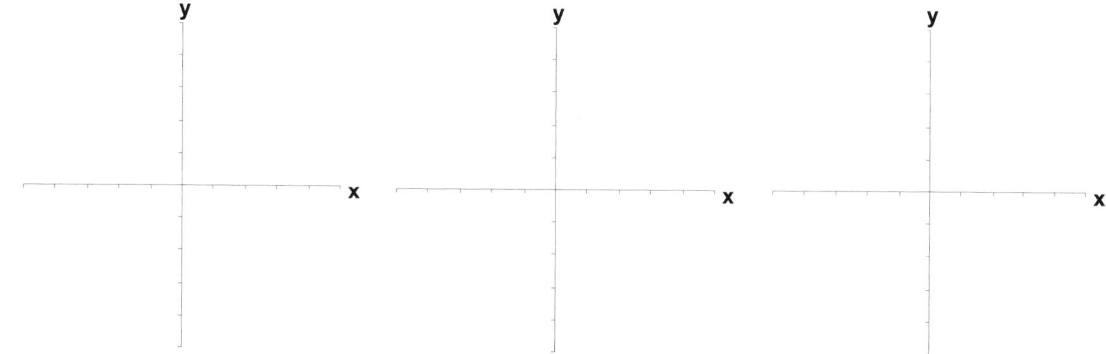

Additional notes

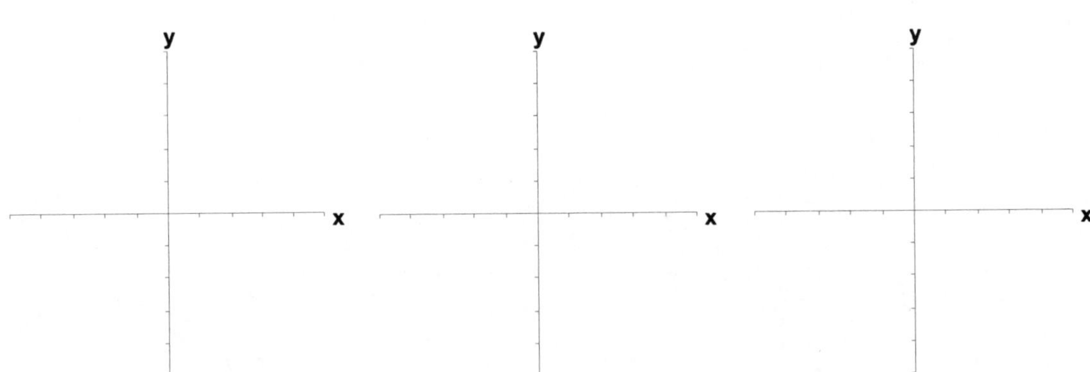

Homework Assignment

Page(s)

Exercises

Name ______________________________ Date __________

Section 7.3 Multivariable Linear Systems

Objective: In this lesson you learned how to solve a system of equations by Gaussian elimination, how to recognize linear systems in row-echelon form and to use back substitution to solve the system, how to solve nonsquare systems of equations, and how to use a system of equations to model and solve real-life problems.

Important Vocabulary Define each term or concept.

Row-echelon form

Gaussian elimination

Nonsquare system of equations

Graph of an equation in three variables

Partial fraction

Partial fraction decomposition

I. Row-Echelon Form and Back-Substitution (Page 489)

What you should learn
How to use back-substitution to solve linear systems in row-echelon form

When elimination is used to solve a system of linear equations, the goal is ______________________________

______________________________.

Example 1: Solve the system of linear equations.

$$\begin{cases} x + y - z = 9 \\ y - 2z = 4 \\ z = 1 \end{cases}$$

II. Gaussian Elimination (Pages 490–492)

What you should learn
How to use Gaussian elimination to solve systems of linear equations

To solve a system that is not in row-echelon form, __________

____

______________________________.

List the three elementary row operations that can be used on a system of linear equations to produce an equivalent system of linear equations.

1.

2.

3.

The number of solution(s) of a system of linear equations in more than two variables must fall into one of the following three categories:

1.

2.

3.

Example 2: Solve the system of linear equations.

$$\begin{cases} x + y + z = 3 \\ 2x - y + 3z = 16 \\ x - 2y - z = 1 \end{cases}$$

A consistent system with exactly one solution is __________________. A consistent system with infinitely many solutions is __________________.

Example 3: The following equivalent system is obtained during the course of Gaussian elimination. Write the solution of the system.

$$\begin{cases} x + 2y - z = 4 \\ \quad y + 2z = 8 \\ \quad\quad 0 = 0 \end{cases}$$

III. Nonsquare Systems (Page 493)

In a square system of linear equations, the number of equations in the system is __________________ the number of variables.

> ***What you should learn***
> How to solve nonsquare systems of linear equations

A system of linear equations cannot have a unique solution unless there are ______________________________

______________________________.

Example 4: Solve the system of linear equations.

$$\begin{cases} x+y+z=1 \\ x-2y-2z=4 \end{cases}$$

IV. Graphical Interpretation of Three-Variable Systems (Page 494)

What you should learn
How to graphically interpret three-variable linear systems

To construct a **three-dimensional coordinate system,** ________

____

______________________________.

To sketch the graph of a plane, ______________________________

____

____

______________________________.

The graph of a system of three linear equations in three variables consists of ______________ planes. When these planes intersect in a single point, the system has ______________ solution(s). When the planes have no point in common, the system has ____________ solution(s). When the planes intersect in a line or a plane, the system has ________________ solution(s).

V. Partial Fraction Decomposition (Pages 495–497)

What you should learn
How to use systems of linear equations to write partial fraction decompositions of rational expressions

Suppose the rational expression $N(x)/D(x)$ is an improper fraction. Before the expression can be decomposed into partial fractions, you must ______________________________

____

____

____

______________________________.

To decompose a proper rational expression into partial fractions, completely factor the denominator into factors of the form

_______________ and ____________________, where

____________________ is irreducible.

Describe how to deal with both linear factors and quadratic factors in the next step of a partial fraction decomposition.

To find the **basic equation** of a partial fraction decomposition, ______________________________________

____.

To solve the basic equation, ____________________________

____

____.

Example 5: Write the form of the partial fraction decomposition for $\dfrac{x-4}{x^2-8x+12}$.

Example 6: Solve the basic equation $5x+3=A(x-1)+B(x+3)$ for A and B.

Homework Assignment
Page(s)
Exercises

Name ____ Date ___________

Section 7.4 Matrices and Systems of Equations

Objective: In this lesson you learned how to write matrices, identify their dimensions, and perform elementary row operations and how to use Gaussian elimination and Gauss-Jordan elimination with matrices to solve systems of linear equations.

Important Vocabulary	Define each term or concept.
Entry of a matrix	
Dimension of a matrix	
Square matrix	
Main diagonal	
Row matrix	
Column matrix	
Elementary row operations	
Gauss-Jordan elimination	

I. Matrices (Pages 504–505)

What you should learn
How to write matrices and identify their dimensions

If m and n are positive integers, define an $m \times n$ **matrix**.

An $m \times n$ matrix has __________ rows and __________ columns.

An **augmented matrix** is ______________________________
____
____.

A **coefficient matrix** is ______________________________

____

____.

Example 1: Consider the following system of equations.

$$\begin{cases} 2x + y - z = 5 \\ x - 3y + 2z = 9 \\ 3x + 2y = 1 \end{cases}$$

(a) Write the augmented matrix for this system.
(b) What is the dimension of the augmented matrix?
(c) Write the coefficient matrix for this system.
(d) What is the dimension of the coefficient matrix?

II. Elementary Row Operations (Page 506)

What you should learn
How to perform elementary row operations on matrices

Two matrices are **row-equivalent** if _______________

____

_______________________________________.

The **elementary row operations** on a matrix are:

III. Gaussian Elimination with Back-Substitution (Pages 507–510)

What you should learn
How to use matrices and Gaussian elimination to solve systems of linear equations

A matrix in **row-echelon form** has the following three properties:
1.

2.

3.

A matrix in row-echelon form is in **reduced row-echelon form** if ____

____.

List the steps for solving a system of linear equations using Gaussian Elimination with Back-Substitution.

If, during the elimination process, you obtain a row with zeros except for the last entry, you can conclude that the system is

____________________________.

Example 2: Solve the following system using Gaussian Elimination with Back-Substitution.

$$\begin{cases} x + y + z = 1 \\ x + 2y + 3z = 1 \\ x - 3y + 5z = -11 \end{cases}$$

IV. Gauss-Jordan Elimination (Pages 511–512)

What you should learn
How to use matrices and Gauss-Jordan elimination to solve systems of linear equations

Example 3: Apply Gauss-Jordan elimination to the following matrix to obtain the unique reduced row-echelon form of the matrix.

$$\left[\begin{array}{ccc:c} 1 & 4 & 2 & 5 \\ 0 & 1 & -1 & 3 \\ 0 & 0 & 1 & -2 \end{array}\right]$$

Example 4: Solve the following system using Gauss-Jordan elimination.

$$\begin{cases} 2x - y + 3z = 1 \\ x + 2y - 4z = -6 \\ -2x + 3y - z = 13 \end{cases}$$

Homework Assignment

Page(s)

Exercises

Name ____ Date ___________

Section 7.5 Operations with Matrices

Objective: In this lesson you learned how to perform operations with matrices.

Important Vocabulary	Define each term or concept.
Scalar multiple	
Zero matrix	
Additive identity	
Matrix multiplication	
Identity matrix of dimension $n \times n$	

I. Equality of Matrices (Page 518)

> ***What you should learn***
> How to decide whether two matrices are equal

Name three ways that a matrix may be represented.

1)

2)

3)

Two matrices are equal if they have the same dimension and ______________________________ are equal.

II. Matrix Addition and Scalar Multiplication (Pages 519–521)

> ***What you should learn***
> How to add and subtract matrices and multiply matrices by scalars

To add two matrices of the same dimension, ______________ ______________________________.

To multiply a matrix A by a scalar c, ______________________ ______________________________.

Example 1: Let $A = \begin{bmatrix} 2 & 5 \\ -3 & 1 \end{bmatrix}$ and $B = \begin{bmatrix} -1 & 4 \\ 2 & -5 \end{bmatrix}$.

Find (a) $A + B$ and (b) $-2B$

Let A, B, and C be $m \times n$ matrices and let c and d be scalars. Give an example of each of the following properties of matrix addition and scalar multiplication:

1) Commutative Property of Matrix Addition: ______________________

2) Associative Property of Matrix Addition: ______________________

3) Associative Property of Scalar Multiplication: ______________________

4) Scalar Identity: ______________________

5) Additive Identity: ______________________

6) Distributive Property (two forms): ______________________

____

III. Matrix Multiplication (Pages 522–524)

> ***What you should learn***
> How to multiply two matrices

The definition of matrix multiplication indicates a row-by-column multiplication, where the entry in the ith row and jth column of the product AB is obtained by ______________________

____

______________________.

Example 2: If A is a 3×5 matrix and B is a 6×3 matrix, find the dimension, if possible, of the product (a) AB, and (b) BA.

Example 3: Find the product AB, if

$$A = \begin{bmatrix} 2 & -1 & 7 \\ 0 & 6 & -3 \end{bmatrix} \quad \text{and} \quad B = \begin{bmatrix} 0 \\ -2 \\ 3 \end{bmatrix}$$

List four properties of Matrix Multiplication:

If A is an $n \times n$ matrix, the identity matrix has the property that

_______________ and _______________.

IV. Applications of Matrix Operations (Pages 525–526)

What you should learn
How to use matrix operations to model and solve real-life problems

Matrix multiplication can be used to represent a system of linear equations. The system

$$\begin{cases} a_{11}x_1 + a_{12}x_2 + a_{13}x_3 = b_1 \\ a_{21}x_1 + a_{22}x_2 + a_{23}x_3 = b_2 \\ a_{31}x_1 + a_{32}x_2 + a_{33}x_3 = b_3 \end{cases}$$

can be written as the matrix equation ____________________, where A is the coefficient matrix of the system and X and B are column matrices.

Example 4: Consider the following system of linear equations.

$$\begin{cases} 2x_1 - x_2 + 3x_3 = -11 \\ \quad x_1 - 3x_3 = -1 \\ -x_1 + 4x_2 + 2x_3 = 2 \end{cases}$$

Write this system as a matrix equation $AX = B$, and then use Gauss-Jordan elimination on the augmented matrix $[A : B]$ to solve for the matrix X.

Additional notes

Homework Assignment

Page(s)

Exercises

Name ____ Date __________

Section 7.6 The Inverse of a Square Matrix

Objective: In this lesson you learned how to find inverses of matrices and how to use inverse matrices to solve systems of linear equations.

Important Vocabulary Define each term or concept.

Inverse of a matrix

I. The Inverse of a Matrix (Page 532)

What you should learn
How to verify that two matrices are inverses of each other

To verify that a matrix B is the inverse of the matrix A, ________ ____________________________________.

II. Finding Inverse Matrices (Pages 533–535)

What you should learn
How to use Gauss-Jordan elimination to find inverses of matrices

If a matrix A has an inverse, A is called ________________ or **nonsingular.** Otherwise, A is called ________________.

A ________________ matrix cannot have an inverse. Not all square matrices have inverses. However, if a matrix does have an inverse, that inverse is ______________.

Describe how to find the inverse of a square matrix A of dimension $n \times n$.

Example 1: Find the inverse of the matrix $A = \begin{bmatrix} 1 & 2 & 4 \\ 1 & 0 & 2 \\ 2 & 3 & 6 \end{bmatrix}$.

III. The Inverse of a 2 × 2 Matrix (Page 536)

> ***What you should learn***
> How to use a formula to find inverses of 2 × 2 matrices

If A is a 2 × 2 matrix given by $A = \begin{bmatrix} a & b \\ c & d \end{bmatrix}$, then A is invertible if and only if ________________. Moreover, if this condition is true, the inverse of A is given by:

$$A^{-1} = \frac{\quad}{\quad}\begin{bmatrix} & \\ & \end{bmatrix}$$

The denominator is called the ____________________ of the 2 × 2 matrix A.

Example 2: Find the inverse of the matrix $B = \begin{bmatrix} 3 & 9 \\ -2 & -7 \end{bmatrix}$.

IV. Systems of Linear Equations (Page 537)

> ***What you should learn***
> How to use inverse matrices to solve systems of linear equations

If A is an invertible matrix, the system of linear equations represented by $AX = B$ has a unique solution given by

________________.

Example 3: Use an inverse matrix to solve (if possible) the system of linear equations:

$$\begin{cases} 12x + 8y = 416 \\ 3x + 5y = 152 \end{cases}$$

> **Homework Assignment**
>
> Page(s)
>
> Exercises

Name ______________________________ Date ________

Section 7.7 The Determinant of a Square Matrix

Objective: In this lesson you learned how to find determinants of square matrices.

Important Vocabulary	Define each term or concept.
Determinant	
Minors	
Cofactors	

I. The Determinant of a 2 × 2 Matrix (Pages 541–542)

What you should learn
How to find the determinants of 2 × 2 matrices

The **determinant** of the 2 × 2 matrix $A = \begin{bmatrix} a_1 & b_1 \\ a_2 & b_2 \end{bmatrix}$ is given by

$$\det(A) = |A| = \begin{vmatrix} \quad & \quad \end{vmatrix} = \_\_\_\_\_\_\_\_\_\_$$

The determinant of a matrix of dimension 1 × 1 is defined as

____.

Example 1: Find the determinant of the matrix $A = \begin{bmatrix} -4 & 3 \\ 1 & -2 \end{bmatrix}$.

II. Minors and Cofactors (Page 543)

What you should learn
How to find minors and cofactors of square matrices

Complete the sign patterns for cofactors of a 3 × 3 matrix, a 4 × 4 matrix, and a 5 × 5 matrix:

Sign Pattern for Cofactors

3 × 3 matrix 4 × 4 matrix 5 × 5 matrix

$$\begin{bmatrix} \quad \end{bmatrix} \qquad \begin{bmatrix} \quad \end{bmatrix} \qquad \begin{bmatrix} \quad \end{bmatrix}$$

Example 2: Use the matrix $A = \begin{bmatrix} 1 & 0 & 3 \\ 2 & 1 & 0 \\ 0 & 2 & 3 \end{bmatrix}$ to find:

(a) the minor M_{13}, and (b) the cofactor C_{21}.

III. The Determinant of a Square Matrix (Page 544)

> ***What you should learn***
> How to find the determinants of square matrices

Applying the definition of the determinant of a square matrix to find a determinant is called ______________________________.

Example 3: Find the determinant of the matrix:

$$A = \begin{bmatrix} -1 & 0 & 4 \\ 3 & -2 & 0 \\ 1 & -1 & 1 \end{bmatrix}$$

Example 4: Describe a strategy for finding the determinant of the following matrix, and then find the determinant of the matrix.

$$B = \begin{bmatrix} -2 & 4 & 0 & 5 \\ 0 & 2 & -1 & 0 \\ 3 & 1 & -4 & -1 \\ -5 & 0 & -2 & 3 \end{bmatrix}$$

Homework Assignment

Page(s)

Exercises

Name ____ Date ____________

Section 7.8 Applications of Matrices and Determinants

Objective: In this lesson you learned how to use Cramer's Rule to solve systems of linear equations.

Important Vocabulary	Define each term or concept.
Uncoded row matrices	
Coded row matrices	

I. Area of a Triangle (Page 548)

> ***What you should learn***
> How to use determinants to find areas of triangles

The area of a triangle with vertices (x_1, y_1), (x_2, y_2), and (x_3, y_3) is

$$\text{Area} = \pm\frac{1}{2}\left|\quad\quad\quad\right|$$

where the symbol ± indicates that the appropriate sign should be chosen to yield a positive area.

Example 1: Find the area of a triangle whose vertices are (– 3, 1), (2, 4), and (5, – 3).

II. Collinear Points (Page 549)

> ***What you should learn***
> How to use determinants to decide whether points are collinear

Collinear points are ________________________________.

Three points (x_1, y_1), (x_2, y_2), and (x_3, y_3) are collinear if and only if

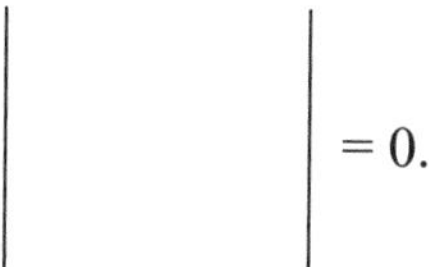

= 0.

Example 2: Determine whether the points (– 2, 4), (0, 3), and (8, – 1) are collinear.

III. Cramer's Rule (Pages 550–552)

What you should learn
How to use Cramer's Rule to solve systems of linear equations

Cramer's Rule states that if a system of n linear equations in n variables has a coefficient matrix A with a nonzero determinant $|A|$, the solution of the system is

$$x_1 = \frac{|A_1|}{|A|}, \quad x_2 = \frac{|A_2|}{|A|}, \ldots, x_n = \frac{|A_n|}{|A|}$$

where the ith column of A_i is ______________________________

____.

If the determinant of the coefficient matrix is ______________,

the system has either no solution or ____________________

______________.

Example 3: Use Cramer's Rule to solve the system of linear equations.

$$\begin{cases} 2x + y + z = 6 \\ -x - y + 3z = 1 \\ y - 2z = -3 \end{cases}$$

IV. Cryptography (Pages 553–555)

What you should learn
How to use matrices to encode and decode messages

A cryptogram is ____________________________________

____________________________.

Describe how to use matrix multiplication to encode and decode messages.

Homework Assignment

Page(s)

Exercises

Name ____ Date __________

Chapter 8 Sequences, Series, and Probability

Section 8.1 Sequences and Series

Objective: In this lesson you learned how to use sequence, factorial, and summation notation to write the terms and sums of sequences.

I. Sequences (Pages 570–572)

What you should learn
How to use sequence notation to write the terms of sequences

An **infinite sequence** is ______________________________
______________________________.

The function values $a_1, a_2, a_3, a_4, \ldots, a_n, \ldots$ are the __________ of an infinite sequence.

A **finite sequence** is ______________________________
______________________________.

To find the first three terms of a sequence, given an expression for its nth term, ______________________________
____
______________________________.

To define a sequence **recursively,** you need to be given ______
______________________________. All other terms of the sequence are then defined using ______________________.

Example 1: Find the first five terms of the sequence given by $a_n = 5 + 2n(-1)^n$.

II. Factorial Notation (Page 573)

What you should learn
How to use factorial notation

If n is a positive integer, ***n*** **factorial** is defined by
____

Zero factorial is defined as ______________________.

Example 2: Evaluate the factorial expression $\dfrac{n!}{(n+1)!}$.

III. Summation Notation (Page 574)

> ***What you should learn***
> How to use summation notation to write sums

The sum of the first n terms of a sequence is represented by the **summation or sigma notation,**

$$\sum_{i=1}^{n} a_i = \_\_\_\_\_\_\_\_\_\_\_\_\_\_\_\_\_\_\_\_\_\_\_\_\_\_\_\_$$

where i is called the __________________________, n is the _____________________________, and 1 is the __________ _______________________.

Example 3: Find the following sum: $\sum_{i=2}^{7}(2+3i)$.

IV. Series (Page 575)

> ***What you should learn***
> How to find sums of infinite series

The sum of the terms of a finite or infinite sequence is called a ____________________.

Consider the infinite sequence $a_1, a_2, a_3, \ldots, a_i, \ldots$. The sum of the first n terms of the sequence is called a(n) _____________________ or the ______________________ of the sequence and is denoted by $a_1 + a_2 + a_3 + \cdots + a_n = \sum_{i=1}^{n} a_i$. The sum of all terms of the infinite sequence is called a(n) _______________________ and is denoted by $a_1 + a_2 + a_3 + \cdots + a_i + \cdots = \sum_{i=1}^{\infty} a_i$.

> **Homework Assignment**
>
> Page(s)
>
> Exercises

Name ____________________ Date __________

Section 8.2 Arithmetic Sequences and Partial Sums

Objective: In this lesson you learned how to recognize, write, and use arithmetic sequences.

Important Vocabulary Define each term or concept.

Arithmetic sequence

I. Arithmetic Sequences (Pages 581–583)

What you should learn
How to recognize, write, and find the nth terms of arithmetic sequences

Define the common difference of an arithmetic sequence.

Example 1: Determine whether or not the following sequence is arithmetic. If it is, find the common difference.
$7, 3, -1, -5, -9, \ldots$

The nth term of an arithmetic sequence has the form

____________________, where d is the common difference between consecutive terms of the sequence, and a_1 is the first term of the sequence.

Example 2: Find a formula for the nth term of the arithmetic sequence whose common difference is 2 and whose first term is 7.

When you know the nth term of an arithmetic sequence *and* you know the common difference of the sequence, you can find the $(n + 1)$th term by using the recursion formula

____________________.

Example 3: Find the sixth term of the arithmetic sequence that begins with 15 and 12.

II. The Sum of a Finite Arithmetic Sequence (Page 584)

> ***What you should learn***
> How to find *n*th partial sums of arithmetic sequences

The sum of a finite arithmetic sequence with *n* terms is given by

____.

The sum of the first *n* terms of an infinite sequence is called the

________________________________.

Example 4: Find the sum of the first 20 terms of the sequence with *n*th term $a_n = 28 - 5n$.

III. Applications (Page 585)

> ***What you should learn***
> How to use arithmetic sequences to model and solve real-life problems

Describe a real-life problem that could be solved by finding the sum of a finite arithmetic sequence.

> **Homework Assignment**
>
> Page(s)
>
> Exercises

Name ______________________________ Date ____________

Section 8.3 Geometric Sequences and Series

Objective: In this lesson you learned how to recognize, write, and use geometric sequences.

Important Vocabulary Define each term or concept.

Geometric sequence

Infinite geometric series or geometric series

I. Geometric Sequences (Pages 589–591)

What you should learn
How to recognize, write, and find the nth terms of geometric sequences

Define the common ratio of a geometric sequence.

Example 1: Determine whether or not the following sequence is geometric. If it is, find the common ratio.
60, 30, 0, – 30, – 60, . . .

The nth term of a geometric sequence has the form

__________________, where r is the common ratio of consecutive terms of the sequence. So, every geometric sequence can be written in the following form:

______________________________________.

If you know the nth term of a geometric sequence, you can find the $(n + 1)$th term by ______________________. That is,

$a_{n+1} =$ ________________.

Example 2: Write the first five terms of the geometric sequence whose first term is $a_1 = 5$ and whose common ratio is – 3.

Example 3: Find the eighth term of the geometric sequence that begins with 15 and 12.

II. The Sum of a Finite Geometric Sequence (Page 592)

> ***What you should learn***
> How to find *n*th partial sums of geometric sequences

The sum of the finite geometric sequence $a_1, a_1r, a_1r^2, a_1r^3, a_1r^4, \ldots, a_1r^{n-1}$ with common ratio $r \neq 1$ is given by

______________________________ .

When using the formula for the sum of a geometric sequence, be careful to check that the index begins with $i = 1$. If the index begins at $i = 0$, ______________________________

______________________________.

Example 4: Find the sum $\sum_{i=1}^{10} 2(0.5)^i$.

III. Geometric Series (Pages 593)

> ***What you should learn***
> How to find sums of infinite geometric series

If $|r| < 1$, then the infinite geometric series $a_1 + a_1r + a_1r^2 + a_1r^3 + a_1r^4 + \ldots + a_1r^{n-1} + \ldots$ has the sum ____________________ .

If $|r| \geq 1$, the series ______________________________ a sum.

Example 5: If possible, find the sum: $\sum_{i=1}^{\infty} 9(0.25)^{i-1}$.

IV. Applications (Page 594)

> ***What you should learn***
> How to use geometric sequences to model and solve real-life problems

Describe a real-life problem that could be solved by finding the sum of a finite geometric sequence.

Homework Assignment

Page(s)

Exercises

Name ______________________________________ Date ____________

Section 8.4 The Binomial Theorem

Objective: In this lesson you learned how to use the Binomial Theorem and Pascal's Triangle to calculate binomial coefficients and write binomial expansions.

Important Vocabulary	Define each term or concept.
Binomial coefficients	
Pascal's Triangle	

I. Binomial Coefficients (Pages 599–600)

> ***What you should learn***
> How to use the Binomial Theorem to calculate binomial coefficients

List four general observations about the expansion of $(x + y)^n$ for various values of n.

1)

2)

3)

4)

The **Binomial Theorem** states that in the expansion of $(x + y)^n = x^n + nx^{n-1}y + \ldots + {}_nC_r x^{n-r}y^r + \ldots + nxy^{n-1} + y^n$, the coefficient of $x^{n-r}y^r$ is ____________________.

Example 1: Find the binomial coefficient ${}_{12}C_5$.

II. Binomial Expansions (Pages 601–602)

> ***What you should learn***
> How to use binomial coefficients to write binomial expansions

Writing out the coefficients for a binomial that is raised to a power is called ____________________.

Example 2: Write the expansion of the expression $(x + 2)^5$.

III. Pascal's Triangle (Page 603)

What you should learn How to use Pascal's Triangle to calculate binomial coefficients

Construct rows 0 through 6 of Pascal's Triangle.

Additional notes

Homework Assignment
Page(s)
Exercises

Name ______________________________ Date ____________

Section 8.5 Counting Principles

Objective: In this lesson you learned how to solve counting problems using the Fundamental Counting Principle, permutations, and combinations.

Important Vocabulary Define each term or concept.

Fundamental Counting Principle

Permutation

Distinguishable permutations

I. Simple Counting Problems (Page 607)

What you should learn
How to solve simple counting problems

If two balls are randomly drawn from a bag of six balls, numbered from 1 to 6, such that it is possible to choose two 3's, the random selection occurs ____________________. If two balls are drawn from the bag at the same time, the random selection occurs ____________________, which eliminates the possibility of choosing two 3's.

II. The Fundamental Counting Principle (Page 608)

What you should learn
How to use the Fundamental Counting Principle to solve more complicated counting problems

The Fundamental Counting Principle can be extended to three or more events. For instance, if E_1 can occur in m_1 ways, E_2 in m_2 ways, and E_3 in m_3 ways, the number of ways that three events E_1, E_2, and E_3 can occur is ____________________.

Example 1: A diner offers breakfast combination plates which can be made from a choice of one of 4 different types of breakfast meats, one of 8 different styles of eggs, and one of 5 different types of breakfast breads. How many different breakfast combination plates are possible?

III. Permutations (Pages 609–611)

> ***What you should learn***
> How to use permutations to solve counting problems

The number of different ways that n elements can be ordered is

______________________________.

A **permutation of n elements taken r at a time** is __________

____

____

____.

The number of n elements taken r at a time is given by

____ .

Example 2: In how many ways can a chairperson, a vice chairperson, and a recording secretary be chosen from a committee of 14 people?

Example 3: In how many distinguishable ways can the letters COMMITTEE be written?

IV. Combinations (Pages 612)

> ***What you should learn***
> How to use combinations to solve counting problems

A **combination of n elements taken r at a time** is __________

____.

The number of combinations of n elements taken r at a time is given by ____ .

Example 4: In how many ways can a research team of 3 students be chosen from a class of 14 students?

> **Homework Assignment**
>
> Page(s)
>
> Exercises

Name ____ Date __________

Section 8.6 Probability

Objective: In this lesson you learned how to find the probability of events.

I. The Probability of an Event (Pages 616–618)

What you should learn
How to find probabilities of events

Any happening whose result is uncertain is called a(n) ________________. The possible results of the experiment are ________________, the set of all possible outcomes of the experiment is the ________________ of the experiment, and any subcollection of a sample space is a(n) ____________.

To calculate the probability of an event, ________________ ____.

If an event E has $n(E)$ equally likely outcomes and its sample space S has $n(S)$ equally likely outcomes, the **probability** of event E is given by ______________________.

The probability of an event must be between _____ and _____.

If $P(E) = 0$, the event E ____________ occur, and E is called a(n) ______________ event. If $P(E) = 1$, the event E ____________ occur, and E is called a(n) ____________ event.

Example 1: A box contains 3 red marbles, 5 black marbles, and 2 yellow marbles. If a marble is selected at random from the box, what is the probability that it is yellow?

II. Mutually Exclusive Events (Pages 619–620)

What you should learn
How to find probabilities of mutually exclusive events

Two events A and B (from the same sample space) are ________ ____________ when A and B have no outcomes in common.

To find the probability that one or the other of two mutually exclusive events will occur, ______________________________ ____.

If A and B are events in the same sample space, the probability of A *or* B occurring is given by $P(A \cup B) =$ ____________________.

If A and B are mutually exclusive, then $P(A \cup B) =$ ____________________.

Example 2: A box contains 3 red marbles, 5 black marbles, and 2 yellow marbles. If a marble is selected at random from the box, what is the probability that it is either red or black?

III. Independent Events (Page 621)

> ***What you should learn***
> How to find probabilities of independent events

Two events are **independent** when ______________________________ ____.

If A and B are **independent events,** the probability that both A and B will occur is $P(A \text{ and } B) =$ ____________________.

That is, to find the probability that two independent events will occur, ____.

Example 3: A box contains 3 red marbles, 5 black marbles, and 2 yellow marbles. If two marbles are randomly selected with replacement, what is the probability that both marbles are yellow?

Homework Assignment
Page(s)

Exercises

Name ____ Date __________

Chapter 9 Topics in Analytic Geometry

Section 9.1 Circles and Parabolas

Objective: In this lesson you learned how to recognize conics, write equations of circles in standard form, write equations of parabolas in standard form, and use the reflective property of parabolas to solve problems.

Important Vocabulary	Define each term or concept.
Directrix	
Focus	
Tangent	

I. Conics (Page 636)

What you should learn
How to recognize a conic as the intersection of a plane and a double-napped cone

A **conic section,** or **conic,** is ____________________________

______________________________________.

Name the four basic conic sections: __________________

____.

In the formation of the four basic conics, the intersecting plane does not pass through the vertex of the cone. When the plane does pass through the vertex, the resulting figure is a(n)

________________________________, such as

____

____

II. Circles (Pages 637–638)

What you should learn
How to write equations of circles in standard form

A **circle** is the set of all points (x, y) in a plane that are

________________________ from a fixed point (h, k), called the

__________________ of the circle. The distance r between the center and any point (x, y) on the circle is the _______________.

The **standard form of the equation of a circle** with center (h, k) and radius r is ________________________.

The standard form of the equation of a circle with radius r and whose center is the origin is ________________________.

Example 1: The point (0, 1) is on a circle whose center is (−2, 1), as shown in the figure. Write the standard form of the equation of the circle.

III. Parabolas (Pages 639–640)

> ***What you should learn***
> How to write equations of parabolas in standard form

A **parabola** is ____
____
____.

The midpoint between the focus and the directrix is the ______________ of a parabola. The line passing through the focus and the vertex is the ______________ of the parabola.

The standard form of the equation of a parabola with a vertical axis having a vertex at (h, k) and directrix $y = k - p$ is

____

The standard form of the equation of a parabola with a horizontal axis having a vertex at (h, k) and directrix $x = h - p$ is

____

The focus lies on the axis p units (directed distance) from the vertex. If the vertex is at the origin (0, 0), the equation takes one of the following forms:

____

Example 2: Find the standard form of the equation of the parabola with vertex at the origin and focus (1, 0).

IV. Reflective Property of Parabolas (Pages 641–642)

> ***What you should learn***
> How to use the reflective property of parabolas to solve real-life problems

Describe a real-life situation in which parabolas are used.

A **focal chord** is ____
____.

The specific focal chord perpendicular to the axis of a parabola is called the ______________________________.

The reflective property of a parabola states that the tangent line to a parabola at a point P makes equal angles with the following two lines:

1)

2)

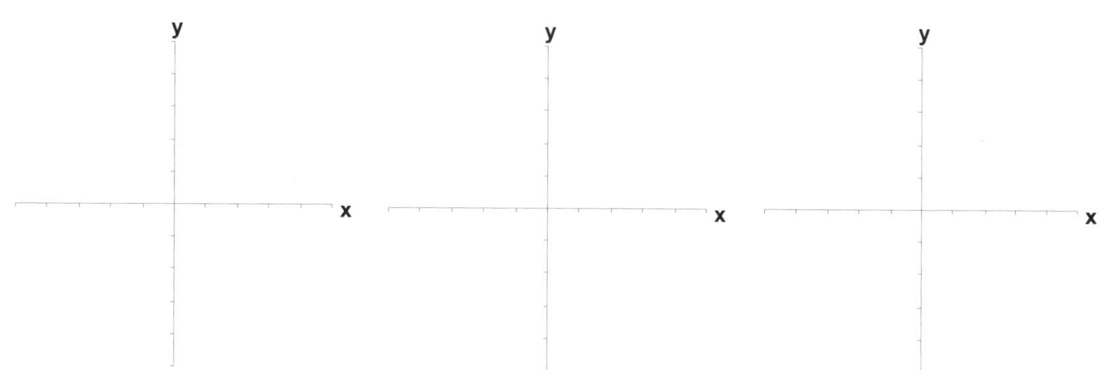

Additional notes

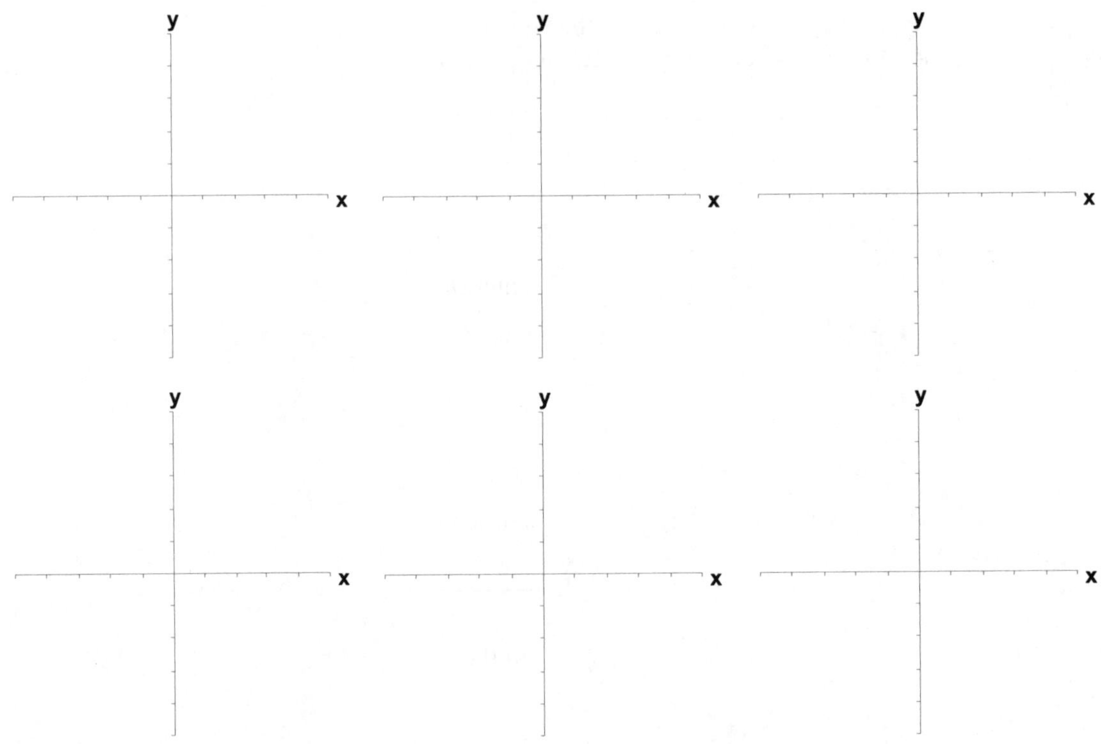

Homework Assignment

Page(s)

Exercises

Name ______________________________ Date ____________

Section 9.2 Ellipses

Objective: In this lesson you learned how to write the standard form of the equation of an ellipse, and analyze and sketch the graphs of ellipses.

Important Vocabulary	Define each term or concept.
Foci	
Vertices	
Major axis	
Center	
Minor axis	

I. Introduction (Pages 647–650)

What you should learn
How to write equations of ellipses in standard form

An **ellipse** is ______________________________
____
______________________________.

The standard form of the equation of an ellipse with center (h, k) and a horizontal major axis of length $2a$ and a minor axis of length $2b$, where $0 < b < a$, is: ______________________________

The standard form of the equation of an ellipse with center (h, k) and a vertical major axis of length $2a$ and a minor axis of length $2b$, where $0 < b < a$, is: ______________________________

In both cases, the foci lie on the major axis, c units from the center, with $c^2 =$ ____________ .

If the center is at the origin (0, 0), the equation takes one of the following forms: ______________________ or ______________________.

Example 1: Sketch the ellipse given by $4x^2 + 25y^2 = 100$.

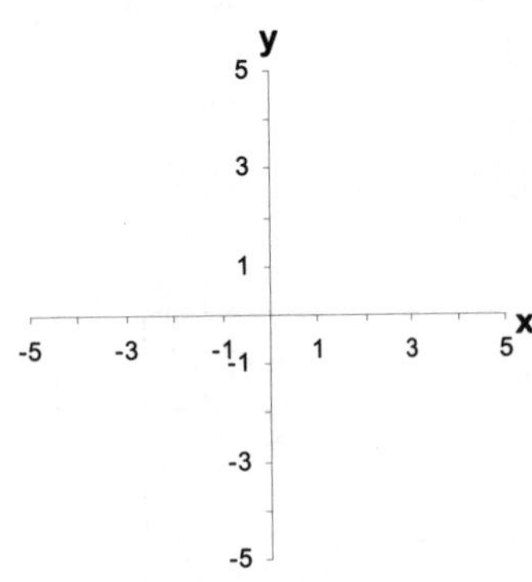

II. Applications (Page 651)

> ***What you should learn***
> How to use properties of ellipses to model and solve real-life problems

Describe a real-life application in which parabolas are used.

III. Eccentricity (Page 652)

> ***What you should learn***
> How to find eccentricities of ellipses

____________________ measures the ovalness of an ellipse. It is given by the ratio $e =$ ____________. For every ellipse, the value of e lies between ________ and ________. For an elongated ellipse, the value of e is close to ____________.

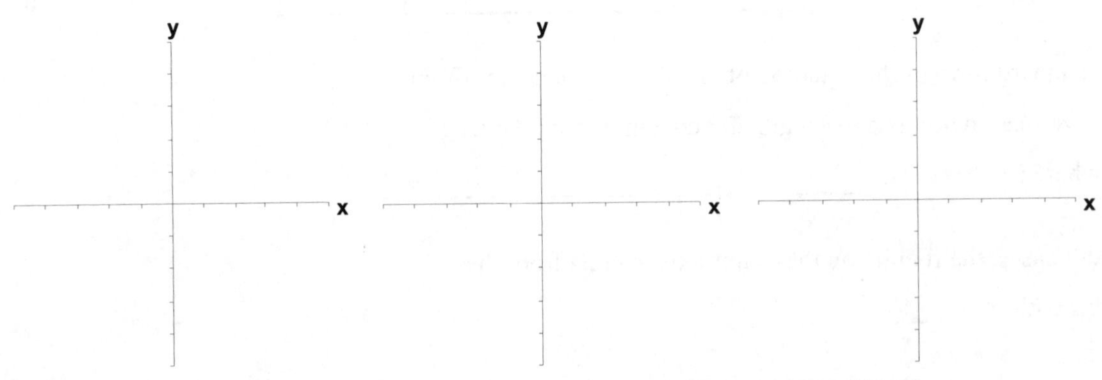

Homework Assignment

Page(s)

Exercises

Name ____ Date __________

Section 9.3 Hyperbolas and Rotation of Conics

Objective: In this lesson you learned how to write the standard form of the equation of a hyperbola, analyze and sketch the graphs of hyperbolas, and rotate axes.

Important Vocabulary Define each term or concept.

Branches

Transverse axis

Conjugate axis

I. Introduction (Pages 656–657)

What you should learn
How to write equations of hyperbolas in standard form

A **hyperbola** is ______________________________________

____

______________________________________.

The line through a hyperbola's two foci intersects the hyperbola at two points called ______________________.

The midpoint of a hyperbola's transverse axis is the ______________ of the hyperbola.

The standard form of the equation of a hyperbola centered at (h, k) and having a horizontal transverse axis is

____

The standard form of the equation of a hyperbola centered at (h, k) and having a vertical transverse axis is

____

In each case, the vertices and foci are, respectively, a and c units from the center. Moreover, a, b, and c are related by the equation

________________.

If the center of the hyperbola is at the origin $(0, 0)$, the equation takes one of the following forms: ____________________ or

____________________.

II. Asymptotes of a Hyperbola (Pages 658–660)

What you should learn
How to find asymptotes of and graph hyperbolas

The **asymptotes** of a hyperbola with a horizontal transverse axis are ______________________.

The **asymptotes** of a hyperbola with a vertical transverse axis are ______________________.

The asymptotes pass through the corners of a rectangle of dimensions ____________ by ____________, with its center at ____________________.

Example 1: Sketch the graph of the hyperbola given by $y^2 - 9x^2 = 9$.

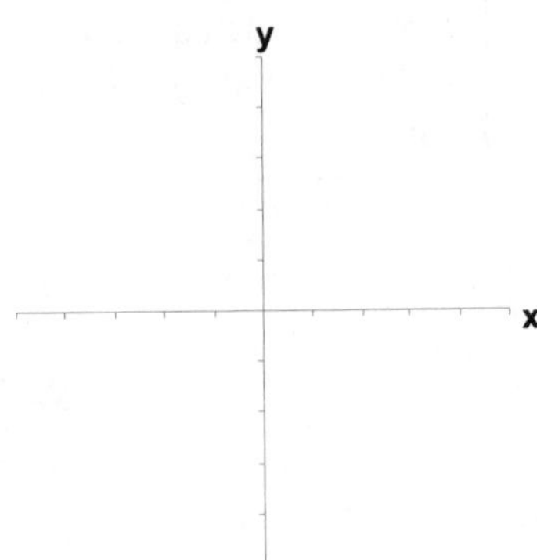

The **eccentricity** of a hyperbola is $e =$ ______________, where the values of e are ______________________.

III. Applications of Hyperbolas (Page 661)

What you should learn
How to use properties of hyperbolas to solve real-life problems

Describe a real-life application in which hyperbolas occur or are used.

IV. General Equations of Conics (Page 662)

> ***What you should learn***
> How to classify conics from their general equations

The graph of $Ax^2 + Bxy + Cy^2 + Dx + Ey + F = 0$ is one of the following:

1) Circle if _______________
2) Parabola if _______________
3) Ellipse if _______________
4) Hyperbola if _______________

Example 2: Classify the equation $9x^2 + y^2 - 18x - 4y + 4 = 0$ as a circle, a parabola, an ellipse, or a hyperbola.

V. Rotation (Pages 663–664)

> ***What you should learn***
> How to rotate the coordinate axes to eliminate the *xy*-term in equations of conics

The general equation of a conic whose axes are rotated so that they are not parallel to either the x-axis or the y-axis contains a(n) _______________.

To eliminate this term, you can use a procedure called ________ _____________, whose objective is to rotate the x- and y-axes until they are parallel to the axes of the conic.

The general second-degree equation

$Ax^2 + Bxy + Cy^2 + Dx + Ey + F = 0$ can be rewritten as

$A'(x')^2 + C'(y')^2 + D'x' + E'y' + F' = 0$ by rotating the coordinate axes through an angle θ, where

$\cot 2\theta =$ _____________________.

The coefficients of the new equation are obtained by making the substitutions $x =$ ________________________________ and $y =$ ________________________________.

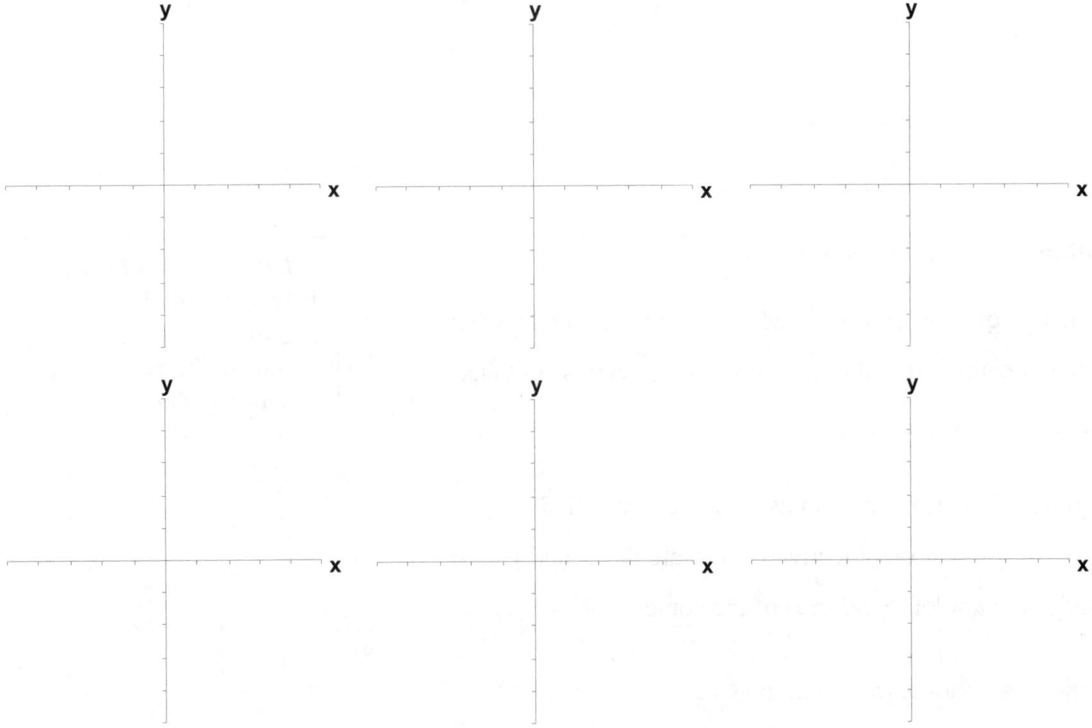

Homework Assignment

Page(s)

Exercises

Name ______________________________ Date __________

Section 9.4 Parametric Equations

Objective: In this lesson you learned how to evaluate sets of parametric equations for given values of the parameter and graph curves that are represented by sets of parametric equations and how to rewrite sets of parametric equations as single rectangular equations and find sets of parametric equations for graphs.

Important Vocabulary Define each term or concept.

Parameter

I. Plane Curves (Page 669)

What you should learn
How to evaluate sets of parametric equations for given values of the parameter

If f and g are continuous functions of t on an interval I, the set of ordered pairs $(f(t), g(t))$ is a(n) ____________________ C. The equations given by $x = f(t)$ and $y = g(t)$ are ________________ ________________ for C, and t is the ________________.

II. Graphs of Plane Curves (Pages 670–671)

What you should learn
How to graph curves that are represented by sets of parametric equations

One way to sketch a curve represented by a pair of parametric equations is to plot points in the ________________. Each set of coordinates (x, y) is determined from a value chosen for the ________________. By plotting the resulting points in the order of increasing values of t, you trace the curve in a specific direction, called the ________________ of the curve.

Example 1: Sketch the curve described by the parametric equations $x = t - 3$ and $y = t^2 + 1$, $-1 \le t \le 3$.

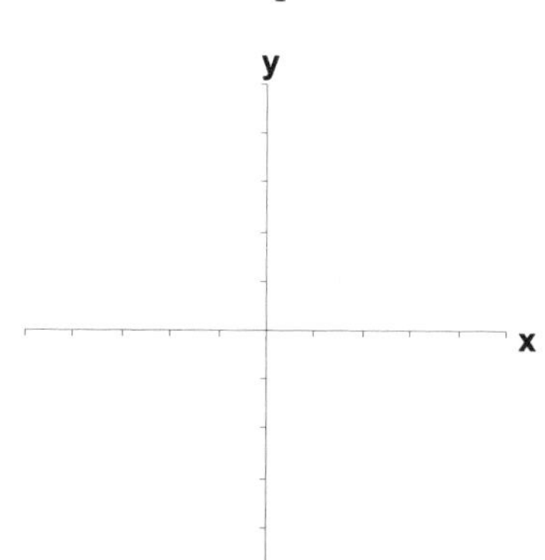

Describe how to display a curve represented by a pair of parametric equations using a graphing utility.

III. Eliminating the Parameter (Pages 672)

> ***What you should learn***
> How to rewrite sets of parametric equations as single rectangular equations by eliminating the parameter

Eliminating the parameter is the process of ________________
____
______________________________________.

Describe the process used to eliminate the parameter from a set of parametric equations.

When converting equations from parametric to rectangular form, it may be necessary to alter ________________________
____
______________________________________.

IV. Finding Parametric Equations for a Graph (Page 673)

> ***What you should learn***
> How to find sets of parametric equations for graphs

Describe how to find a set of parametric equations for a given graph.

> **Homework Assignment**
>
> Page(s)
>
> Exercises

Name ____ Date __________

Section 9.5 Polar Coordinates

Objective: In this lesson you learned how to plot points in the polar coordinate system and convert equations from rectangular to polar form and vice versa.

I. Introduction (Pages 677–678)

> ***What you should learn***
> How to plot points and find multiple representations of points in the polar coordinate system

To form the **polar coordinate system** in the plane, fix a point O, called the ______________ or ______________, and construct from O an initial ray called the ________________. Then each point P in the plane can be assigned ________________ ________ as follows:

1) $r =$ ________________________________

2) $\theta =$ ____

____

In the polar coordinate system, points do not have a unique representation. For instance, the point (r, θ) can be represented as ____________________ or ____________________, where n is any integer. Moreover, the pole is represented by $(0, \theta)$, where θ is ____________________.

Example 1: Plot the point $(r, \theta) = (-2, 11\pi/4)$ on the polar coordinate system.

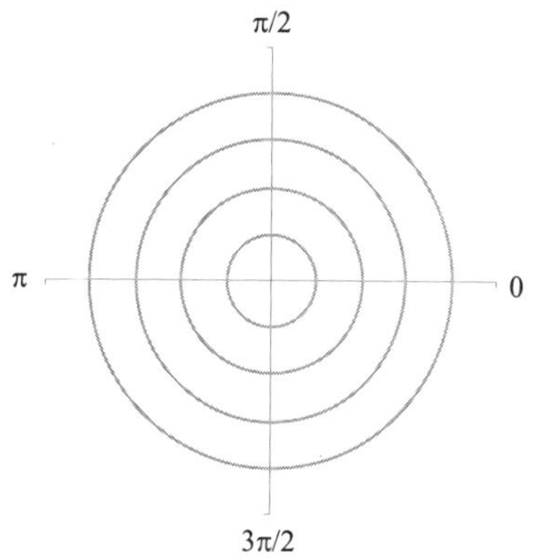

Example 2: Find another polar representation of the point $(4, \pi/6)$.

II. Coordinate Conversion (Page 679)

> ***What you should learn***
> How to convert points from rectangular to polar form and vice versa

The polar coordinates (r, θ) are related to the rectangular coordinates (x, y) as follows:

Example 3: Convert the polar coordinates $(3, 3\pi/2)$ to rectangular coordinates.

III. Equation Conversion (Page 680)

> ***What you should learn***
> How to convert equations from rectangular to polar form and vice versa

To convert a rectangular equation to polar form, ____________

____.

Example 4: Find the rectangular equation corresponding to the polar equation $r = \dfrac{-5}{\sin\theta}$.

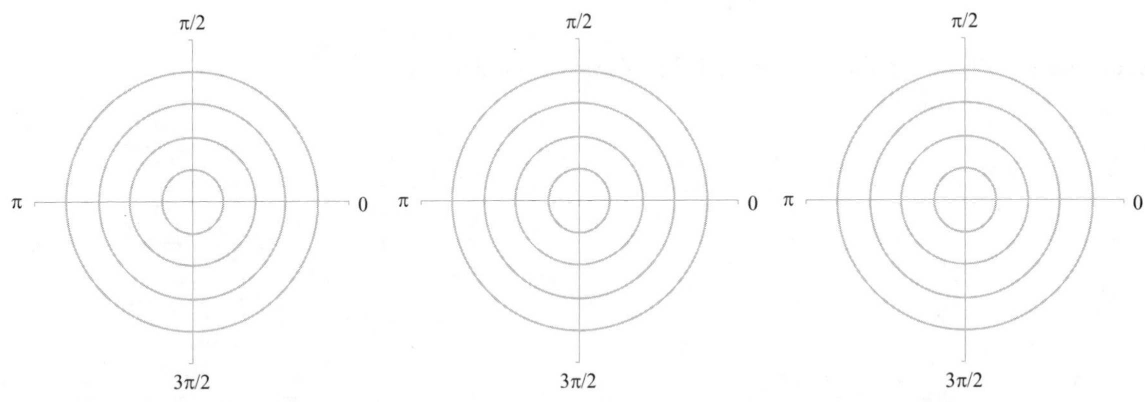

Homework Assignment

Page(s)

Exercises

Name ____ Date ___________

Section 9.6 Graphs of Polar Equations

Objective: In this lesson you learned how to graph polar equations.

I. Introduction (Page 683)

> ***What you should learn***
> How to graph polar equations by point plotting

Example 1: Use point plotting to sketch the graph of the polar equation $r = 3\cos\theta$.

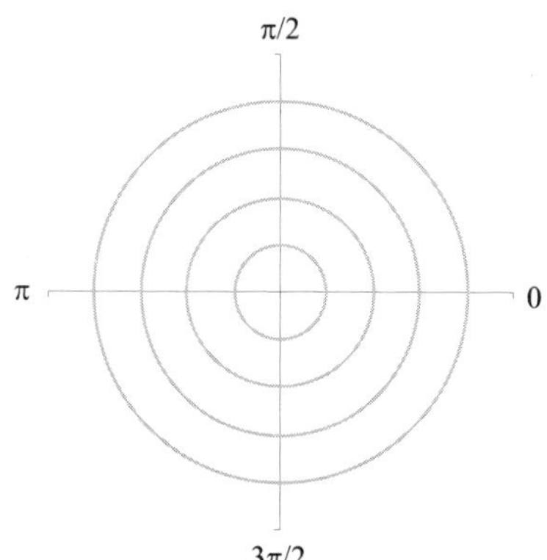

The graph of the polar equation $r = f(\theta)$ can be rewritten in parametric form, using *t* as a parameter, as follows:

____

II. Symmetry and Zeros (Pages 684–686)

> ***What you should learn***
> How to use symmetry and zeros as sketching aids

The graph of a polar equation is symmetric with respect to the following if the given substitution yields an equivalent equation.

Substitution

1) The line $\theta = \pi/2$:

2) The polar axis:

3) The pole:

Example 2: Describe the symmetry of the polar equation $r = 2(1 - \sin\theta)$.

An additional aid to sketching graphs of polar equations is

____.

Example 3: Describe the zeros of the polar equation
$r = 5\cos 2\theta$

IV. Special Polar Graphs (Pages 687–688)

What you should learn
How to recognize special polar graphs

List the general equations that yield each of the following types of special polar graphs:

Limaçons:

Rose curves:

Circles:

Lemniscates:

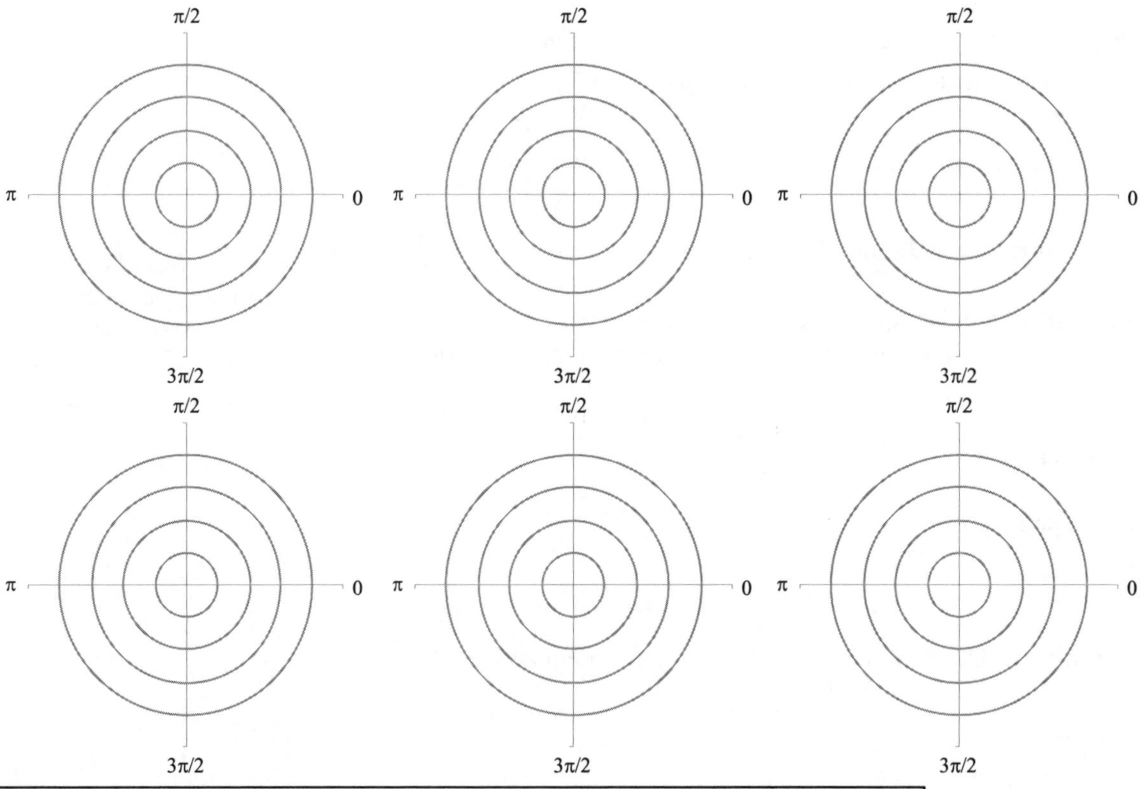

Homework Assignment

Page(s)

Exercises

Name ____ Date ___________

Section 9.7 Polar Equations of Conics

Objective: In this lesson you learned how to write conics in terms of eccentricity and to write equations of conics in polar form.

I. Alternative Definition of Conics and Polar Equations (Pages 691–693)

> ***What you should learn***
> How to define conics in terms of eccentricities and write and graph equations of conics in polar form

The locus of a point in the plane that moves so that its distance from a fixed point (focus) is in a constant ratio to its distance from a fixed line (directrix) is a ______________. The constant ratio is the ______________________ of the conic and is denoted by e. Moreover, the conic is an ellipse if ___________, a parabola if ___________, and a hyperbola if ____________.

For each type of conic, the focus is at the _____________.

The graph of the polar equation ________________________ is a conic with a vertical directrix to the right of the pole, where $e > 0$ is the eccentricity and $|p|$ is the distance between the focus (pole) and the directrix.

The graph of the polar equation ________________________ is a conic with a vertical directrix to the left of the pole, where $e > 0$ is the eccentricity and $|p|$ is the distance between the focus (pole) and the directrix.

The graph of the polar equation ________________________ is a conic with a horizontal directrix above the pole, where $e > 0$ is the eccentricity and $|p|$ is the distance between the focus (pole) and the directrix.

The graph of the polar equation ________________________ is a conic with a horizontal directrix below the pole, where $e > 0$ is the eccentricity and $|p|$ is the distance between the focus (pole) and the directrix.

Example 1: Identify the type of conic from the polar equation $r = \dfrac{36}{10 + 12\sin\theta}$, and describe its orientation.

II. Applications (Page 694)

> ***What you should learn***
> How to use equations of conics in polar form to model real-life problems

Describe a real-life application of polar equations of conics.

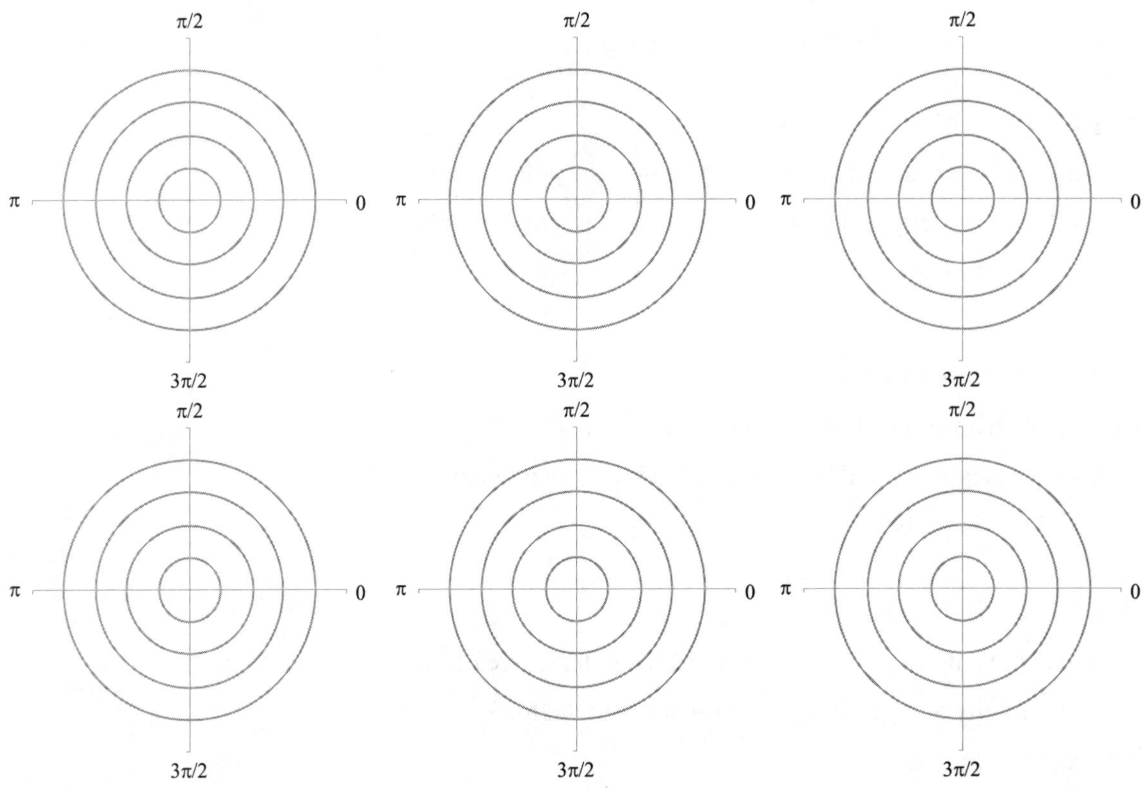

Homework Assignment

Page(s)

Exercises

Name ____ Date ____________

Chapter 10 Analytic Geometry in Three Dimensions

Section 10.1 The Three-Dimensional Coordinate System

Objective: In this lesson you learned how to plot points, find distances between points, and find midpoints of line segments connecting points in space and how to write equations of spheres and graph traces of surfaces in space.

Important Vocabulary Define each term or concept.

Solid analytic geometry

Sphere

Surface in space

Trace

I. The Three-Dimensional Coordinate System (Page 712)

What you should learn
How to plot points in the three-dimensional coordinate system

A **three-dimensional coordinate system** is constructed by

____

____.

Taken as pairs, the axes determine three **coordinate planes:** the

______________, the ______________, and the ______________.

These three coordinate planes separate the three-dimensional coordinate system into eight ______________. The first octant is the one in which ______________________________

______________________________.

In the three-dimensional system, a point P in space is determined by an ordered triple (x, y, z), where x, y, and z are as follows.

$x =$ ________________________________

$y =$ ________________________________

$z =$ ________________________________

II. The Distance and Midpoint Formulas (Page 713)

> ***What you should learn***
> How to find distances between points in space and find midpoints of line segments joining points in space

The distance between the points (x_1, y_1, z_1) and (x_2, y_2, z_2) given by the **Distance Formula in Space** is

The midpoint of the line segment joining the points (x_1, y_1, z_1) and (x_2, y_2, z_2) given by the **Midpoint Formula in Space** is

$$\left(\qquad \qquad \qquad \right)$$

Example 1: For the points (2, 0, – 4) and (– 1, 4, 6), find
(a) the distance between the two points, and
(b) the midpoint of the line segment joining them.

III. The Equation of a Sphere (Pages 714–715)

> ***What you should learn***
> How to write equations of spheres in standard form and find traces of surfaces in space

The standard equation of a sphere whose center is (h, k, j) and whose radius is r is ____________________.

Example 2: Find the center and radius of the sphere whose equation is $x^2 + y^2 + z^2 - 4x + 2y + 8z + 17 = 0$.

To find the yz-trace of a surface, ____________________

____

____

____

____________________.

> **Homework Assignment**
>
> Page(s)
>
> Exercises

Name ______________________________ Date ____________

Section 10.2 Vectors in Space

Objective: In this lesson you learned how to represent vectors and find dot products of and angles between vectors in space.

Important Vocabulary Define each term or concept.

Standard unit vector notation in space

Angle between two nonzero vectors in space

Parallel vectors in space

I. Vectors in Space (Pages 719–720)

What you should learn
How to find the component forms of, the unit vectors in the same direction of, the magnitudes of, the dot products of, and the angles between vectors in space

In space, vectors are denoted by ordered triples of the form

______________________ .

The **zero vector in space** is denoted by ________________.

If **v** is represented by the directed line segment from $P(p_1, p_2, p_3)$ to $Q(q_1, q_2, q_3)$, describe how the **component form** of **v** is produced.

Two vectors are **equal** if and only if ____________________

____.

The **magnitude**(or length) of $\mathbf{u} = \langle u_1, u_2, u_3 \rangle$ is:

$\| \mathbf{u} \| = \sqrt{}$

A unit vector **u** in the direction of **v** is ________________.

The **sum** of $\mathbf{u} = \langle u_1, u_2, u_3 \rangle$ and $\mathbf{v} = \langle v_1, v_2, v_3 \rangle$ is

$\mathbf{u} + \mathbf{v} =$ ______________________ .

The **scalar multiple** of the real number c and $\mathbf{u} = \langle u_1, u_2, u_3 \rangle$ is

$c\mathbf{u} =$ ______________________ .

The **dot product** of $\mathbf{u} = \langle u_1, u_2, u_3 \rangle$ and $\mathbf{v} = \langle v_1, v_2, v_3 \rangle$ is

$\mathbf{u} \bullet \mathbf{v} =$ ____________________________

If θ is the angle between two nonzero vectors **u** and **v**, then

$\cos \theta =$ ___________________________________.

If the dot product of two nonzero vectors is zero, the angle between the vectors is ____________. Such vectors are called

____________________.

Example 1: Find the dot product of the vectors $\langle -1, 4, -2 \rangle$ and $\langle 0, -1, 5 \rangle$.

II. Parallel Vectors (Pages 721–722)

> ***What you should learn***
> How to determine whether vectors in space are parallel or orthogonal

Example 2: Determine whether the vectors $\langle 6, 1, -3 \rangle$ and $\langle -2, -1/3, 1 \rangle$ are parallel.

To use vectors to determine whether three points P, Q, and R in space are collinear, ___________________________________

____.

III. Application (Page 723)

> ***What you should learn***
> How to use vectors in space to solve real-life problems

Describe a real-life application of vectors in space.

> **Homework Assignment**
>
> Page(s)
>
> Exercises

Name ____ Date __________

Section 10.3 The Cross Product of Two Vectors

Objective: In this lesson you learned how to find cross products of vectors in space, use geometric properties of the cross product, and use triple scalar products to find volumes of parallelepipeds.

I. The Cross Product (Pages 726–727)

> ***What you should learn***
> How to find cross products of vectors in space

A vector in space that is orthogonal to two given vectors is called their ____________________.

Let $\mathbf{u} = u_1\mathbf{i} + u_2\mathbf{j} + u_3\mathbf{k}$ and $\mathbf{v} = v_1\mathbf{i} + v_2\mathbf{j} + v_3\mathbf{k}$ be two vectors in space. The **cross product** of **u** and **v** is the vector

$\mathbf{u} \times \mathbf{v} =$ ____

Describe a convenient way to remember the formula for the cross product.

Example 1: Given $\mathbf{u} = -2\mathbf{i} + 3\mathbf{j} - 3\mathbf{k}$ and $\mathbf{v} = \mathbf{i} - 2\mathbf{j} + \mathbf{k}$, find the cross product $\mathbf{u} \times \mathbf{v}$.

Let **u**, **v**, and **w** be vectors in space and let c be a scalar. Complete the following properties of the cross product:

1. $\mathbf{u} \times \mathbf{v} =$ ________________
2. $\mathbf{u} \times (\mathbf{v} + \mathbf{w}) =$ ____________________
3. $c(\mathbf{u} \times \mathbf{v}) =$ ______________________
4. $\mathbf{u} \times \mathbf{0} =$ ________________
5. $\mathbf{u} \times \mathbf{u} =$ __________
6. $\mathbf{u} \bullet (\mathbf{v} \times \mathbf{w}) =$ ________________

II. Geometric Properties of the Cross Product (Pages 728–729)

> ***What you should learn***
> How to use geometric properties of cross products of vectors in space

Complete the following geometric properties of the cross product, given **u** and **v** are nonzero vectors in space and θ is the angle between **u** and **v**.

1. $\mathbf{u} \times \mathbf{v}$ is orthogonal to ____________________.

2. $\|\mathbf{u} \times \mathbf{v}\|$ = ____________________.

3. $\mathbf{u} \times \mathbf{v} = \mathbf{0}$ if and only if ____________________.

4. $\|\mathbf{u} \times \mathbf{v}\|$ = area of the parallelogram having ____________________ ____________________.

III. The Triple Scalar Product (Page 730)

> ***What you should learn***
> How to use triple scalar products to find volumes of parallelepipeds

For vectors **u**, **v**, and **w** in space, the dot product of **u** and $\mathbf{v} \times \mathbf{w}$ is called the ____________________ of **u**, **v**, and **w**, and is found as

$$\mathbf{u} \bullet (\mathbf{v} \times \mathbf{w}) = \left| \quad \right|$$

The volume V of a parallelepiped with vectors **u**, **v**, and **w** as adjacent edges is ____________________.

Example 2: Find the volume of the parallelepiped having $\mathbf{u} = 2\mathbf{i} + \mathbf{j} - 3\mathbf{k}$, $\mathbf{v} = \mathbf{i} - 2\mathbf{j} + 3\mathbf{k}$, and $\mathbf{w} = 4\mathbf{i} - 3\mathbf{k}$ as adjacent edges.

Homework Assignment

Page(s)

Exercises

Name ______________________________ Date ____________

Section 10.4 Lines and Planes in Space

Objective: In this lesson you learned how to find parametric and symmetric equations of lines in space and find distances between points and planes in space.

I. Lines in Space (Pages 733–734)

> ***What you should learn***
> How to find parametric and symmetric equations of lines in space

For the line L through the point $P = (x_1, y_1, z_1)$ and parallel to the vector $\mathbf{v} = \langle a, b, c \rangle$, the vector v is the ____________________ for the line L, and the values a, b, and c are the ____________ ____________.

One way of describing the line L is ____________________ ____ ____.

A line L parallel to the nonzero vector $\mathbf{v} = \langle a, b, c \rangle$ and passing through the point $P = (x_1, y_1, z_1)$ is represented by the following parametric equations, where t is the parameter:

____

If the direction numbers a, b, and c are all nonzero, you can eliminate the parameter t to obtain the **symmetric equations** of a line:

____

II. Planes in Space (Pages 735–737)

> ***What you should learn***
> How to find equations of planes in space

The plane containing the point (x_1, y_1, z_1) and having nonzero normal vector $\mathbf{n} = \langle a, b, c \rangle$ can be represented by the **standard form of the equation of a plane,** which is

____

By regrouping terms, you obtain the **general form of the equation of a plane** in space:

____

To find a normal vector to a plane given the general form of the equation of the plane, ______________________________ ____.

Two distinct planes in three-space either are ________________

or ____________________________.

If two distinct planes intersect, the **angle θ between the two planes** is equal to the angle between vectors $\mathbf{n}_1$ and $\mathbf{n}_2$ normal to the two intersecting planes, and is given by

____

Consequently, two planes with normal vectors $\mathbf{n}_1$ and $\mathbf{n}_2$ are

1. ____________________ if $\mathbf{n}_1 \bullet \mathbf{n}_2 = 0$.

2. ___________________ if $\mathbf{n}_1$ is a scalar multiple of $\mathbf{n}_2$.

III. Sketching Planes in Space (Page 738)

What you should learn
How to sketch planes in space

If a plane in space intersects one of the coordinate planes, the line of intersection is called the ______________ of the given plane in the coordinate plane.

To sketch a plane in space, ______________________________

____

____.

The plane with equation $3y - 2z + 1 = 0$ is parallel to

_____________________.

IV. Distance Between a Point and a Plane (Page 739)

What you should learn
How to find distances between points and planes in space

The **distance between a plane and a point** Q (not in the plane) is

____

where P is a point in the plane and $\mathbf{n}$ is normal to the plane.

Homework Assignment

Page(s)

Exercises

Name ______________________________ Date __________

Chapter 11 Limits and an Introduction to Calculus

Section 11.1 Introduction to Limits

Objective: In this lesson you learned how to estimate limits and use properties and operations of limits.

I. The Limit Concept and Definition of Limit (Pages 750–752)

What you should learn
How to understand the limit concept and use the definition of a limit to estimate limits

Define **limit.**

Describe how to estimate the limit $\lim_{x \to -2} \frac{x^2 + 4x + 4}{x + 2}$ numerically.

The existence or nonexistence of $f(x)$ when $x = c$ has no bearing on the existence of ______________________.

II. Limits That Fail to Exist (Pages 753–754)

What you should learn
How to determine whether limits of functions exist

The limit of $f(x)$ as $x \to c$ does not exist under any of the following conditions.

1.

2.

3.

Give an example of a limit that does not exist.

III. Properties of Limits and Direct Substitution (Pages 755–756)

> ***What you should learn***
> How to use properties of limits and direct substitution to evaluate limits

Let b and c be real numbers and let n be a positive integer. Complete each of the following properties of limits.

1. $\lim_{x\to c} b =$ ____________
2. $\lim_{x\to c} x =$ ____________
3. $\lim_{x\to c} x^n =$ ____________
4. $\lim_{x\to c} \sqrt[n]{x} =$ ______________________________

Let b and c be real numbers, let n be a positive integer, and let f and g be functions with the following limits.

$$\lim_{x\to c} f(x) = L \quad \text{and} \quad \lim_{x\to c} g(x) = K$$

Complete each of the following statements about operations with limits.

1. Scalar multiple: $\lim_{x\to c}[b\,f(x)] =$ ____________
2. Sum or difference: $\lim_{x\to c}[f(x) \pm g(x)] =$ ______________
3. Product: $\lim_{x\to c}[f(x) \cdot g(x)] =$ ____________
4. Quotient: $\lim_{x\to c} \dfrac{f(x)}{g(x)} =$ ______________________
5. Power: $\lim_{x\to c}[f(x)]^n =$ ____________

Example 1: Find the limit: $\lim_{x\to 4} 3x^2$.

If p is a polynomial function and c is a real number, then

$\lim_{x \to c} p(x) =$ ______________.

If r is a rational function given by $r(x) = p(x)/q(x)$, and c is a real number such that $q(c) \neq 0$, then

$\lim_{x \to c} r(x) =$ ___________________________.

Example 2: Find the limit: $\lim_{x \to 2} \dfrac{4 - x^2}{x}$.

Additional notes

Additional notes

Homework Assignment

Page(s)

Exercises

Name ____ Date ___________

Section 11.2 Techniques for Evaluating Limits

Objective: In this lesson you learned how to find limits by direct substitution and by using the dividing out and rationalizing techniques.

I. Dividing Out Technique (Pages 760–761)

> *What you should learn*
> How to use the dividing out technique to evaluate limits of functions

The validity of the **dividing out technique** stems from ________

____

____.

The dividing out technique should be applied only when

____

____.

An **indeterminate form** is ____________________________

____

____

____

____.

When you encounter an indeterminate form by direct substitution into a rational function, you can conclude ________________

____.

Example 1: Find the following limit: $\lim_{x \to 3} \frac{x^2 - 8x + 15}{x - 3}$.

II. Rationalizing Technique (Page 762)

> *What you should learn*
> How to use the rationalizing technique to evaluate limits of functions

Another way to find the limits of some functions is to first rationalize the numerator. This is called the ______________ ____________, which means multiplying the numerator and denominator by the conjugate of the numerator.

III. Using Technology (Page 763)

To find limits of nonalgebraic functions, ________________

____

____

____

____.

> ***What you should learn***
> How to use technology to approximate limits of functions graphically and numerically

IV. One-Sided Limits (Pages 764–765)

Define a **one-sided limit**.

> ***What you should learn***
> How to evaluate one-sided limits of functions

Existence of a Limit
If f is a function and c and L are real numbers, then $\lim_{x \to c} f(x) = L$

if and only if ______________________________________

____.

V. A Limit from Calculus (Page 766)

For any x-value, the limit of a *difference quotient* is an expression of the form

> ***What you should learn***
> How to evaluate limits of difference quotients from calculus

____

Direct substitution into the difference quotient always produces

____.

Homework Assignment

Page(s)

Exercises

Name ______________________________ Date __________

Section 11.3 The Tangent Line Problem

Objective: In this lesson you learned how to approximate slopes of tangent lines, use the limit definition of slope, and use derivatives to find slopes of graphs.

I. Tangent Line to a Graph (Page 770)

> ***What you should learn***
> How to understand the tangent line problem

The **tangent line** to the graph of a function f at a point $P(x_1, y_1)$ is ____

______________________________.

To determine the rate at which a graph rises or falls at a single point, ____

______________________________.

II. Slope of a Graph (Page 771)

> ***What you should learn***
> How to use a tangent line to approximate the slope of a graph at a point

To visually approximate the slope of a graph at a point, _______

____

____

____

______________________________.

III. Slope and the Limit Process (Pages 772–774)

> ***What you should learn***
> How to use the limit definition of slope to find exact slopes of graphs

A **secant line** to a graph is ______________________________

____.

A **difference quotient** is ______________________________.

Give the definition of the slope of a graph.

Example 1: Use the limit process to find the slope of the graph of $f(x) = x^2 + 5$ at the point (3, – 1).

IV. The Derivative of a Function (Pages 775–776)

What you should learn
How to find derivatives of functions and use derivatives to find slopes of graphs

The derivative of f at x is the function derived from ____.

Give the formal definition of the **derivative.**

The derivative $f'(x)$ is a formula for ____.

Example 2: Find the derivative of $f(x) = 9 - 2x^2$.

y x y x y x

Homework Assignment

Page(s)

Exercises

Name ____ Date __________

Section 11.4 Limits at Infinity and Limits of Sequences

Objective: In this lesson you learned how to evaluate limits at infinity and find limits of sequences.

I. Limits at Infinity and Horizontal Asymptotes (Pages 780–783)

> ***What you should learn***
> How to evaluate limits of functions at infinity

Define **limits at infinity.**

To help evaluate limits at infinity, you can use the following:

If r is a positive real number, then $\lim_{x\to\infty} \frac{1}{x^r} =$ ___________.

If x^r is defined when $x < 0$, then $\lim_{x\to-\infty} \frac{1}{x^r} =$ ___________.

Example 1: Find the limit: $\lim_{x\to\infty} \frac{1+5x-3x^3}{x^3}$

If $f(x)$ is a rational function and the limit of f is taken as x approaches ∞ or $-\infty$,

- When the degree of the numerator is less than the degree of the denominator, the limit is ________________.
- When the degrees of the numerator and the denominator are equal, the limit is ________________________ ________________________.
- When the degree of the numerator is greater than the degree of the denominator, the limit ______________ ______________________.

II. Limits of Sequences (Pages 784–785)

What you should learn
How to find limits of sequences

For a sequence whose nth term is a_n, as n increases without bound, if the terms of the sequence get closer and closer to a particular value L, then the sequence is said to _______________ to L. Otherwise, a sequence that does not converge is said to _______________.

Give the definition of the limit of a sequence.

Example 2: Find the limit of the sequence $a_n = \dfrac{(n-3)(4n-1)}{4-3n-n^2}$.

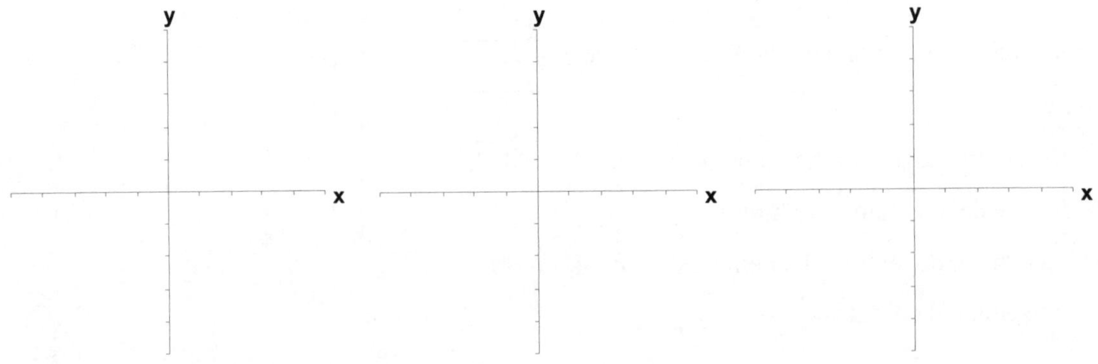

Homework Assignment

Page(s)

Exercises

Name ____ Date ____________

Section 11.5 The Area Problem

Objective: In this lesson you learned how to find limits of summations and use them to find areas of regions bounded by graphs of functions.

I. Limits of Summations (Pages 789–791)

What you should learn
How to find limits of summations

The following summation formulas and properties are used to evaluate finite and infinite summations.

1. $\sum_{i=1}^{n} c =$ ____________

2. $\sum_{i=1}^{n} i =$ ____________________

3. $\sum_{i=1}^{n} i^2 =$ __________________________

4. $\sum_{i=1}^{n} i^3 =$ ____________________

5. $\sum_{i=1}^{n} (a_i \pm b_i) = \sum_{i=1}^{n} a_i \pm \sum_{i=1}^{n} b_i$

6. $\sum_{i=1}^{n} ka_i = k\sum_{i=1}^{n} a_i$

To find the limit of a summation, _________________________
____
____
____.

Example 1: Find the limit of $S(n)$ as $n \to \infty$.

$$S(n) = \sum_{i=1}^{n} \frac{i-5}{n^3}$$

II. The Area Problem (Pages 792–793)

> ***What you should learn***
> How to use rectangles to approximate and limits of summations to find areas of plane regions

Describe the area problem.

The exact **area of a plane region *R*** is given by __________
____.

Let f be continuous and nonnegative on the interval $[a, b]$. The area A of the region bounded by the graph of f, the x-axis, and the vertical lines $x = a$ and $x = b$ is given by

____

Example 2: Find the area of the region bounded by the graph of $f(x) = (x - 4)^2 + 5$ and the x-axis between $x = 3$ and $x = 6$.

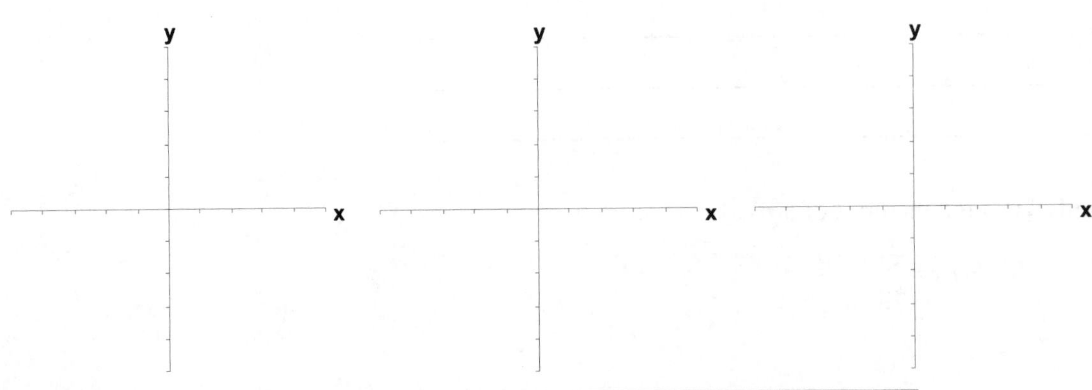

> **Homework Assignment**
>
> Page(s)
>
> Exercises

Name ____ Date __________

Appendix C Review of Graphs, Equations, and Inequalities

Section C.1 The Cartesian Plane

Objective: In this lesson you learned how to plot points in the coordinate plane and use the Distance and Midpoint Formulas.

Important Vocabulary	Define each term or concept.
Rectangular coordinate system, or Cartesian plane	

I. The Cartesian Plane (Pages A46–A47)

> ***What you should learn***
> How to plot points in the Cartesian plane

On the Cartesian plane, the horizontal real number line is usually called the ______________, and the vertical real number line is usually called the ______________. The origin is the ________ __________________ of these two axes, and the two axes divide the plane into four parts called ________________.

On the Cartesian plane shown below, label the x-axis, the y-axis, the origin, Quadrant I, Quadrant II, Quadrant III, and Quadrant IV.

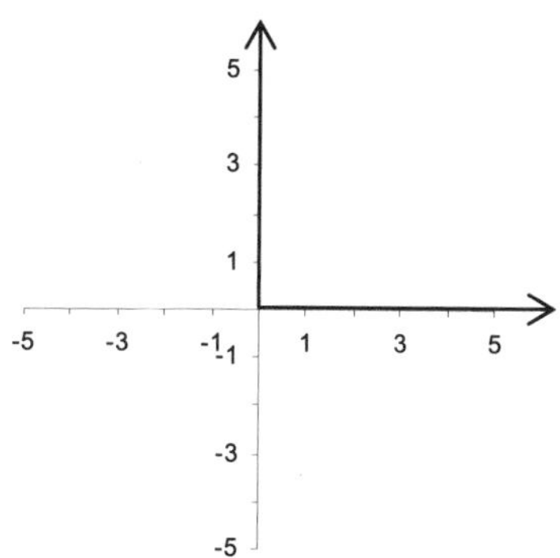

Each point in the plane corresponds to an **ordered pair** (x, y) of real numbers x and y, called __________________________ of the point. The ***x*-coordinate** represents the directed distance from ____, and the ***y*-coordinate** represents ______________________________ ____.

Example 1: Explain how to plot the ordered pair (3, – 2), and then plot it on the Cartesian plane provided on the next page.

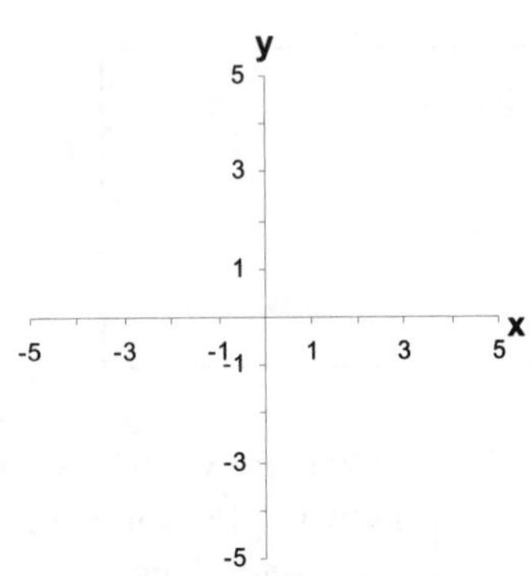

II. The Distance Formula (Pages A47–A48)

What you should learn
How to use the Distance Formula to find the distance between two points

The **Distance Formula** states that ____________________

____

____

Example 2: Explain how to use the Distance Formula to find the distance between the points (4, 2) and (5, – 1). Then find the distance and round to the nearest hundredth.

Example 3: Explain how to use a graphical solution to find the distance between the points (4, 2) and (5, – 1).

III. The Midpoint Formula (Page A49)

> ***What you should learn***
> How to use the Midpoint Formula to find the midpoint of a line segment

To find the **midpoint** of a line segment that joins two points in a coordinate plane, ____

____

______________________________.

The **Midpoint Formula** gives the midpoint of the line segment joining the points (x_1, y_1) and (x_2, y_2) as

____

Example 4: Explain how to find the midpoint of the line segment with endpoints at (– 8, 2) and (6, – 10). Then find the coordinates of the midpoint.

IV. The Equation of a Circle (Page A50)

> ***What you should learn***
> How to find the equation of a circle

A **circle** in the plane consists of ______________________________

____.

The **standard form of the equation of a circle** with center (h, k) and radius r is ____________________.

The standard form of the equation of a circle with radius r and whose center is at the origin is ____________________.

Example 5: For the equation $(x+2)^2 + (y-1)^2 = 4$, find the center and radius of the circle and then sketch the graph of the equation.

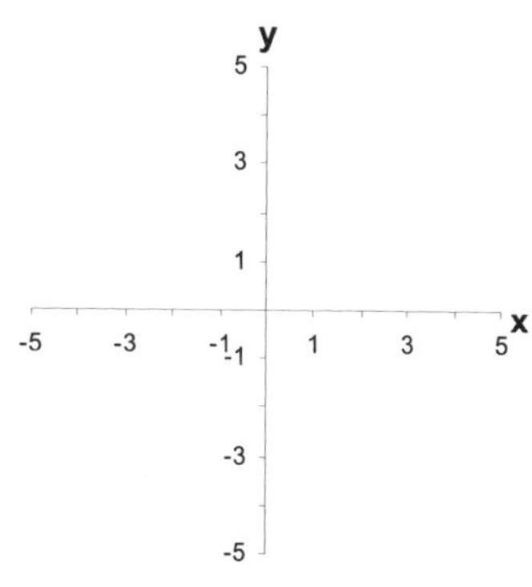

V. Application (Page A51)

> ***What you should learn***
> How to translate points in the plane

To shift a figure plotted in the rectangular coordinate system by a units to the left and b units upward, ____________________

____

____.

If (x, y) is an original point on a graph, __________________ is a reflection of this original point in the y-axis. If (x, y) is an original point on a graph, __________________ is a reflection of the original point in the x-axis. If (x, y) is an original point on a graph, __________________ is a reflection of the original point through the origin.

Additional notes

y x y x y x

Homework Assignment

Page(s)

Exercises

Name ____ Date ___________

Section C.2 Graphs of Equations

Objective: In this lesson you learned how to sketch graphs of equations by point plotting or using a graphing utility.

Important Vocabulary	Define each term or concept.
Solution point	
Graph of an equation	
Intercepts	

I. The Graph of an Equation (Pages A55–A56)

What you should learn
How to sketch graphs of equations by point plotting

List the steps for sketching the graph of an equation by point plotting:

Example 1: Complete the table for the equation $y = 3 - 0.5x$. Then use point plotting to sketch the graph of the equation.

x	-4	-2	0	2	4
y					

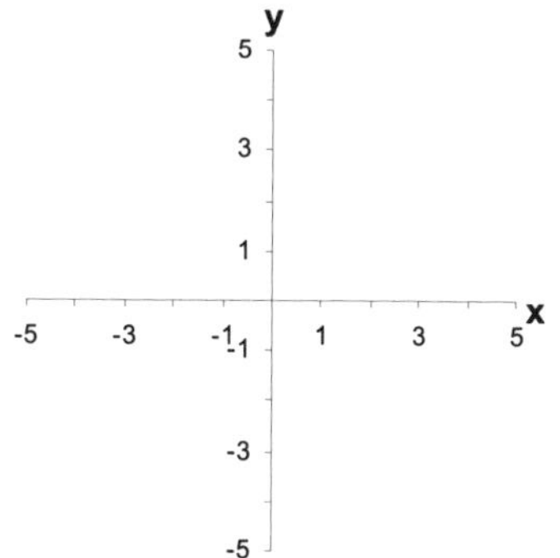

II. Using a Graphing Utility (Pages A57–A58)

> ***What you should learn***
> How to graph equations using a graphing utility

A disadvantage of the point-plotting method is ______________

____

____

____.

To graph an equation involving x and y on a graphing utility:

Example 2: Use a graphing utility to graph the equation $12x^2 + 4y = 5$ in a standard viewing window.

A square setting is ______________________________

____

____.

A square setting is useful when using a graphing utility to graph _____________________________________.

Example 3: Describe how to use a graphing utility to graph $3x^2 + 3y^2 = 75$. Then graph the equation in a square viewing window.

III. Applications of Graphs of Equations (Pages A59–A60)

What you should learn
How to use graphs of equations to solve real-life problems

List and describe three common approaches to solving a problem.

1)

2)

3)

Describe a real-life situation in which a graphical solution approach would be helpful.

Example 4: Suppose a toy company estimates that its top-selling toy sells 240 units per minute, on average, nationally during the holiday shopping season, or according to the equation $S = 240m$, where S is the number of units sold and m is the number of minutes. Explain how a graphing utility could be used to find how long it takes during the holiday shopping season to sell 82,800 units.

Additional notes

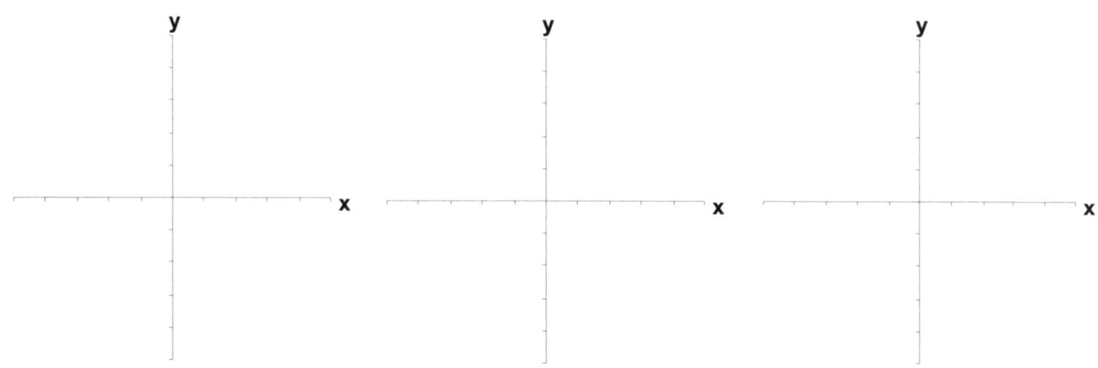

Additional notes

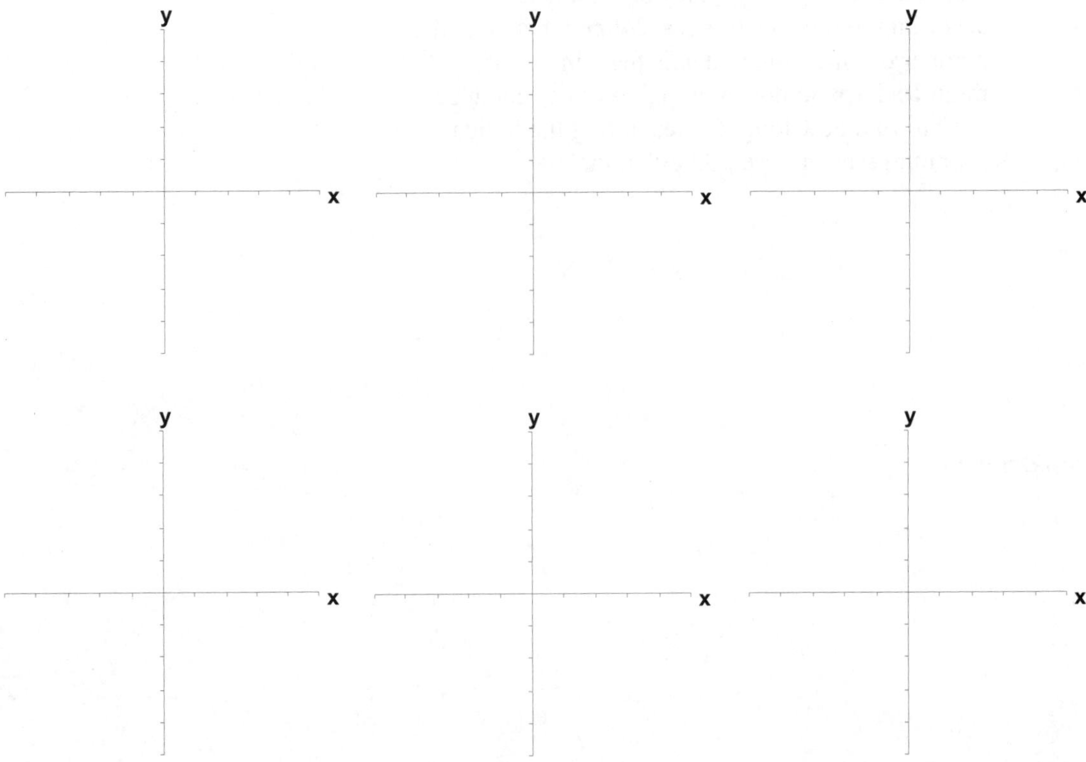

Homework Assignment

Page(s)

Exercises

Name ____ Date __________

Section C.3 Solving Equations Algebraically and Graphically

Objective: In this lesson you learned how to solve linear equations, find solutions of equations graphically, find the points of intersection of two graphs, and solve polynomial equations as well as those involving radicals, fractions, or absolute values.

Important Vocabulary	Define each term or concept.
Equation	
Extraneous solution	
***x*-intercept**	
***y*-intercept**	

I. Equations and Solutions of Equations (Pages A64–A65)

> ***What you should learn***
> How to solve linear equations

To solve an equation in x means to ______________________

____.

The values of x for which the equation is true are called its

__________________.

An identity equation is ______________________________

____.

A conditional equation is ______________________________

____

____.

A **linear equation in one variable x** is an equation that can be written in the standard form ________________, where a and b are real numbers, with $a \neq$ _______.

Example 1: Solve $5(x+3)=35$.

Explain how to solve an equation involving fractional expressions.

When is it possible to introduce an extraneous solution?

Example 2: Solve: (a) $\frac{5x}{7}=\frac{9}{14}$ (b) $\frac{1}{x+1}+\frac{5x}{x^2-1}=\frac{4}{x-1}$

II. Intercepts and Solutions (Pages A65–A67)

What you should learn
How to find x- and y-intercepts of graphs of equations

To find the x-intercepts of the graph of an equation, __________

____.

To find the y-intercepts of the graph of an equation, __________

____.

Example 3: For the equation $3x-4y=12$, find:
(a) the x-intercept(s), and (b) the y-intercept(s).

III. Finding Solutions Graphically (Pages A67–A68)

What you should learn
How to find solutions of equations graphically

Describe how to use a graphing utility to graphically approximate the solutions of an equation.

Example 4: Use a graphing utility to approximate the solutions of $3x^2 - 14x = -8$.

IV. Points of Intersection of Two Graphs (Pages A69–A70)

> ***What you should learn***
> How to find the points of intersection of two graphs

Describe how to find the points of intersection of the graphs of two equations algebraically.

Describe how to approximate the points of intersection of the graphs of two equations with a graphing utility.

Example 5: Use (a) an algebraic approach and (b) a graphical approach to finding the points of intersection of the graphs of $y = 2x^2 - 5x + 6$ and $x - y = -6$.

V. Solving Polynomial Equations Algebraically (Pages A71–A72)

> ***What you should learn***
> How to solve polynomial equations

List four methods for solving quadratic equations:

Describe how to solve a quadratic equation by factoring.

Example 6: Solve $x^2 - 12x = -27$ by factoring.

List the solutions given by the Quadratic Formula to solve the quadratic equation written in general form as $ax^2 + bx + c = 0$.

Example 7: Describe a strategy for solving the polynomial equation $x^3 + 2x^2 - x = 2$. Then find the solutions.

An equation is of **quadratic type** if ____________________________

____.

VI. Equations Involving Radicals (Page A73)

What you should learn
How to solve equations involving radicals

An equation involving a radical expression can often be cleared of radicals by ____

____.

When using this procedure, remember to check for ____________

____________________________, which do not satisfy the original equation.

Example 8: Describe a strategy for solving the following equation involving a radical expression: $\sqrt{8-x}-15=0$

VII. Equations Involving Fractions or Absolute Values (Pages A74–A75)

What you should learn
How to solve equations involving fractions or absolute values

Describe how to solve an equation involving fractions.

Example 9: Solve: $\frac{2}{x}-1=\frac{1}{x+1}$

Describe how to solve an equation involving an absolute value.

Example 10: Write the two equations that must be solved to solve the absolute value equation $\left|3x^2+2x\right|-5=0$.

Additional notes

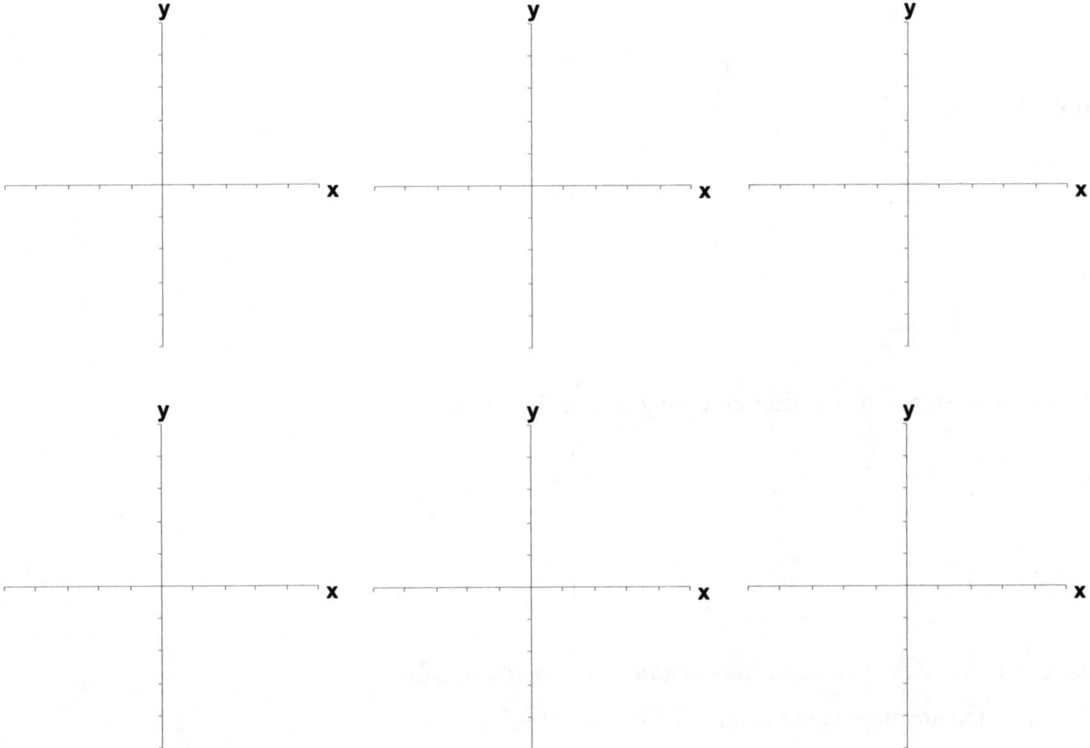

Homework Assignment

Page(s)

Exercises

Name ____ Date ____________

Section C.4 Solving Inequalities Algebraically and Graphically

Objective: In this lesson you learned how to solve linear inequalities, inequalities involving absolute values, polynomial inequalities, and rational inequalities.

Important Vocabulary Define each term or concept.

Solutions of an inequality

Graph of an inequality

Double inequality

Critical numbers

Test intervals

I. Properties of Inequalities (Pages A80–A82)

What you should learn
How to use properties of inequalities to solve linear inequalities

Solving an inequality in the variable x means ______________
____.

Values that are solutions of an inequality are said to
__________________ the inequality.

To solve a linear inequality in one variable, use the __________
____________________ to isolate the variable.

When each side of an inequality is multiplied or divided by a negative number ______________________________________
____.

Two inequalities that have the same solution set are
____.

Complete the list of Properties of Inequalities given below.

1) Transitive Property: $a < b$ and $b < c$ $\rightarrow$ ______________
2) Addition of Inequalities: $a < b$ and $c < d$ $\rightarrow$ ______________
3) Addition of a Constant c: $a < b$ $\rightarrow$ ______________

4) Multiplication by a Constant c:

For $c > 0$, $a < b \rightarrow$ ____________________

For $c < 0$, $a < b \rightarrow$ ____________________

Describe the steps that would be necessary to solve the linear inequality $7x - 2 < 9x + 8$.

Describe how to use a graphing utility to solve the linear inequality $7x - 2 < 9x + 8$.

The two inequalities $-10 < 3x$ and $14 \geq 3x$ can be rewritten as the double inequality ________________________.

II. Inequalities Involving Absolute Value (Page A83)

What you should learn
How to solve inequalities involving absolute values

Let x be a variable or an algebraic expression and let a be a real number such that $a \geq 0$. The solutions of $|x| < a$ are all values of x that ____________________________________. The solutions of $|x| > a$ are all values of x that _________________________ ________________________.

Example 1: Solve the inequality: $|x + 11| - 4 \leq 0$

The symbol $\cup$ is called a ______________ symbol and is used to denote ____.

Example 2: Write the following solution set using interval notation: $x > 8$ or $x < 2$

III. Polynomial Inequalities (Pages A84–A86)

> ***What you should learn***
> How to solve polynomial inequalities

Where can polynomials change signs?

Between two consecutive zeros, a polynomial must be

____.

When the real zeros of a polynomial are put in order, they divide the real number line into ____________________________

____.

These zeros are the ____________________ of the inequality, and the resulting open intervals are the ____________________.

Complete the following steps for determining the intervals on which the values of a polynomial are entirely negative or entirely positive:

1)

2)

3)

Explain how to approximate the solution of the polynomial inequality $3x^2 + 2x - 5 < 0$ from a graph.

If a polynomial inequality is not given in general form, you should begin the solution process by ______________________

____

____.

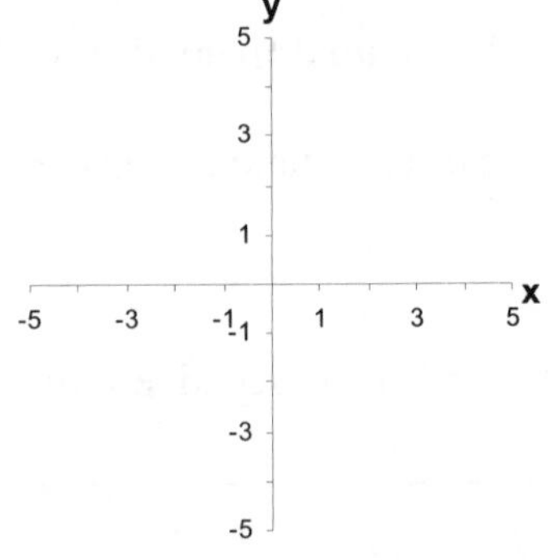

Example 3: Solve $x^2 + x - 20 \geq 0$.

Example 4: Use a graph to solve the polynomial inequality $-x^2 - 6x - 9 > 0$.

IV. Rational Inequalities (Page A87)

What you should learn
How to solve rational inequalities

To extend the concepts of key numbers and test intervals to rational inequalities, use the fact that the value of a rational expression can change sign only at its _______________ and its _________________________. These two types of numbers make up the ______________________ of a rational inequality.

Describe how to solve a rational inequality.

Example 5: Solve $\dfrac{3x+15}{x-2} \leq 0$.

Homework Assignment

Page(s)

Exercises